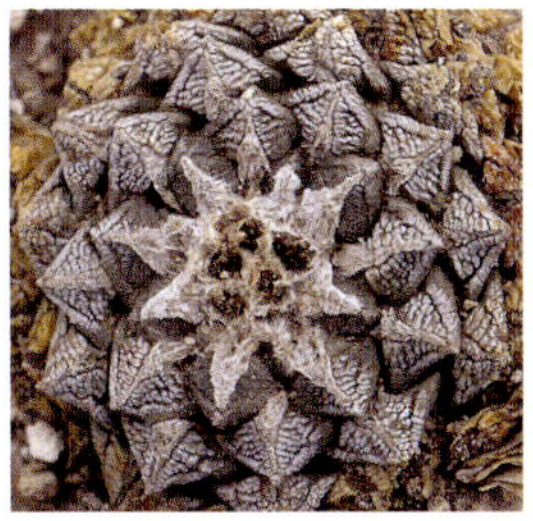

Magic resides within this small canyon, the equal of the most spectacular in America. With the early morning sunrise, viewed from the top of Dalquest peak, one is mesmerized by the interwoven formations of light and color on the landscape. The intermingled softness of green-golden vegetation against the stark reality of desert stone and the bright colors of flowers where water abounds create a new awareness of a desert environment known only to a few. The bloom of the ocotillo, the flower of the yucca, the contrasts of earth and sky, wind and erosion, the changes of seasons, all have contributed to an unbelievable experience of living in this canyon. This is a rugged, spellbinding canyon, enclosing delicate life forms and minute natural patterns within vast, overpowering monuments of stone. The canyon is full of stories for those who know how to read them.

—*Walter W. Nelson*

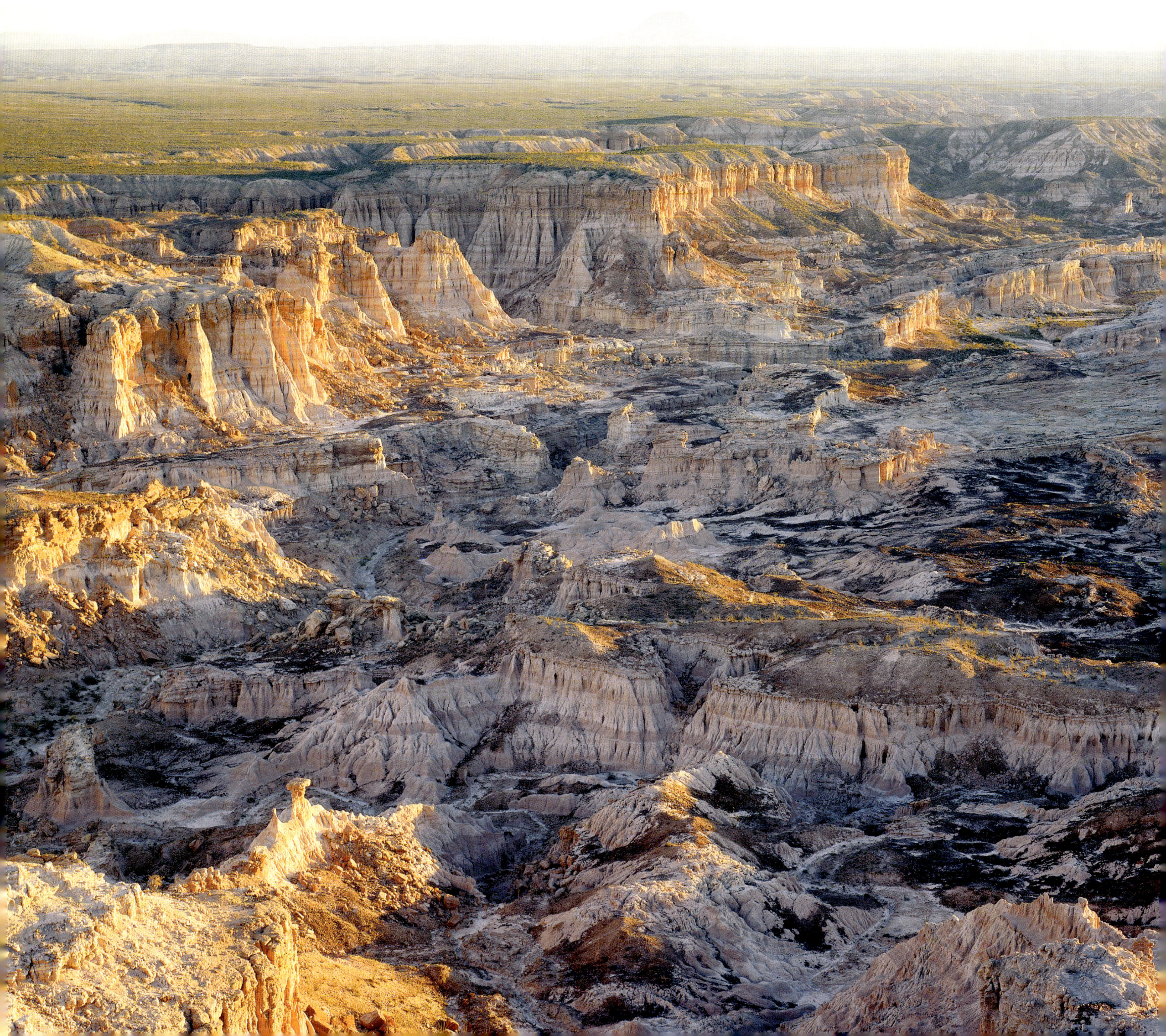

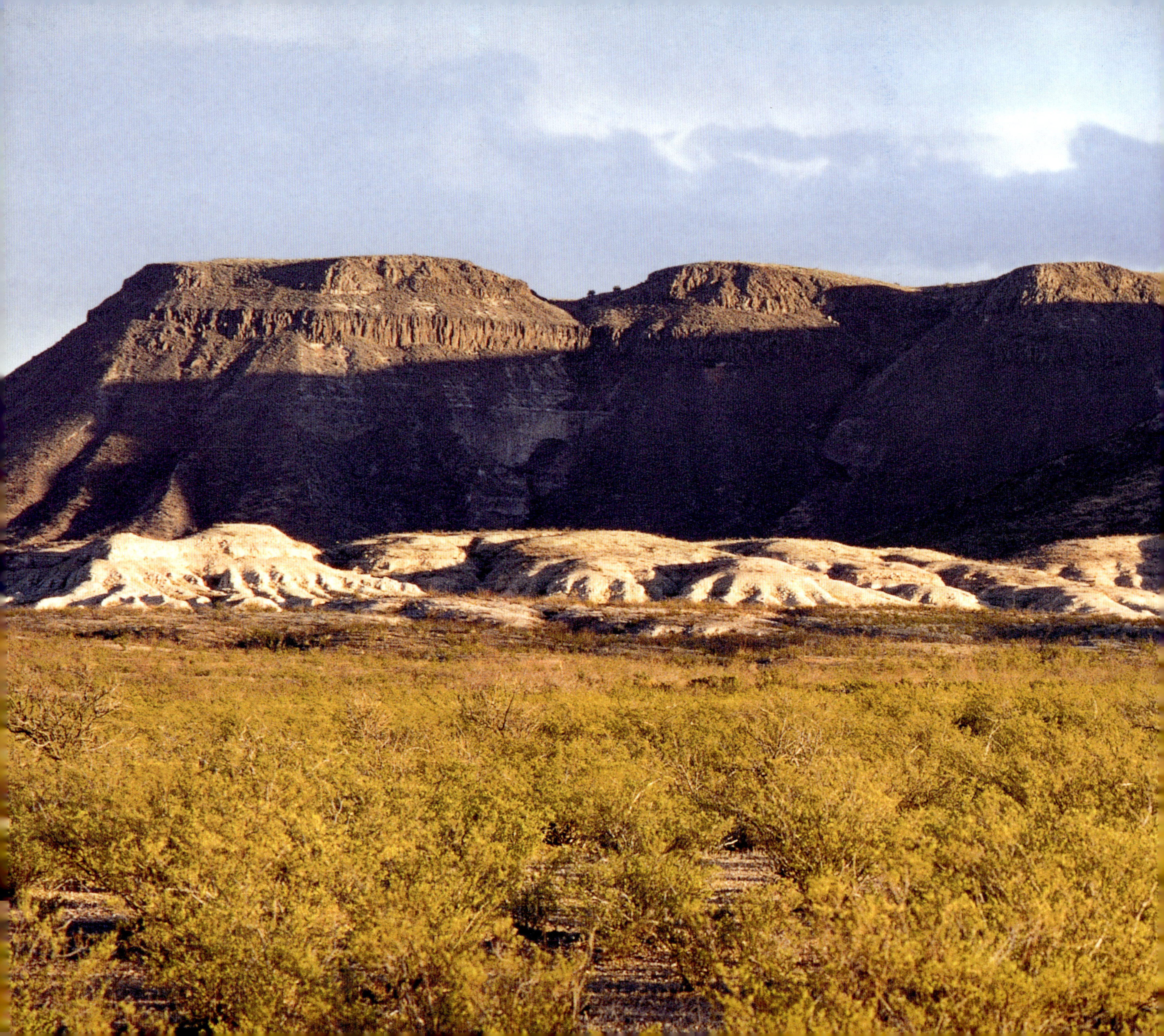

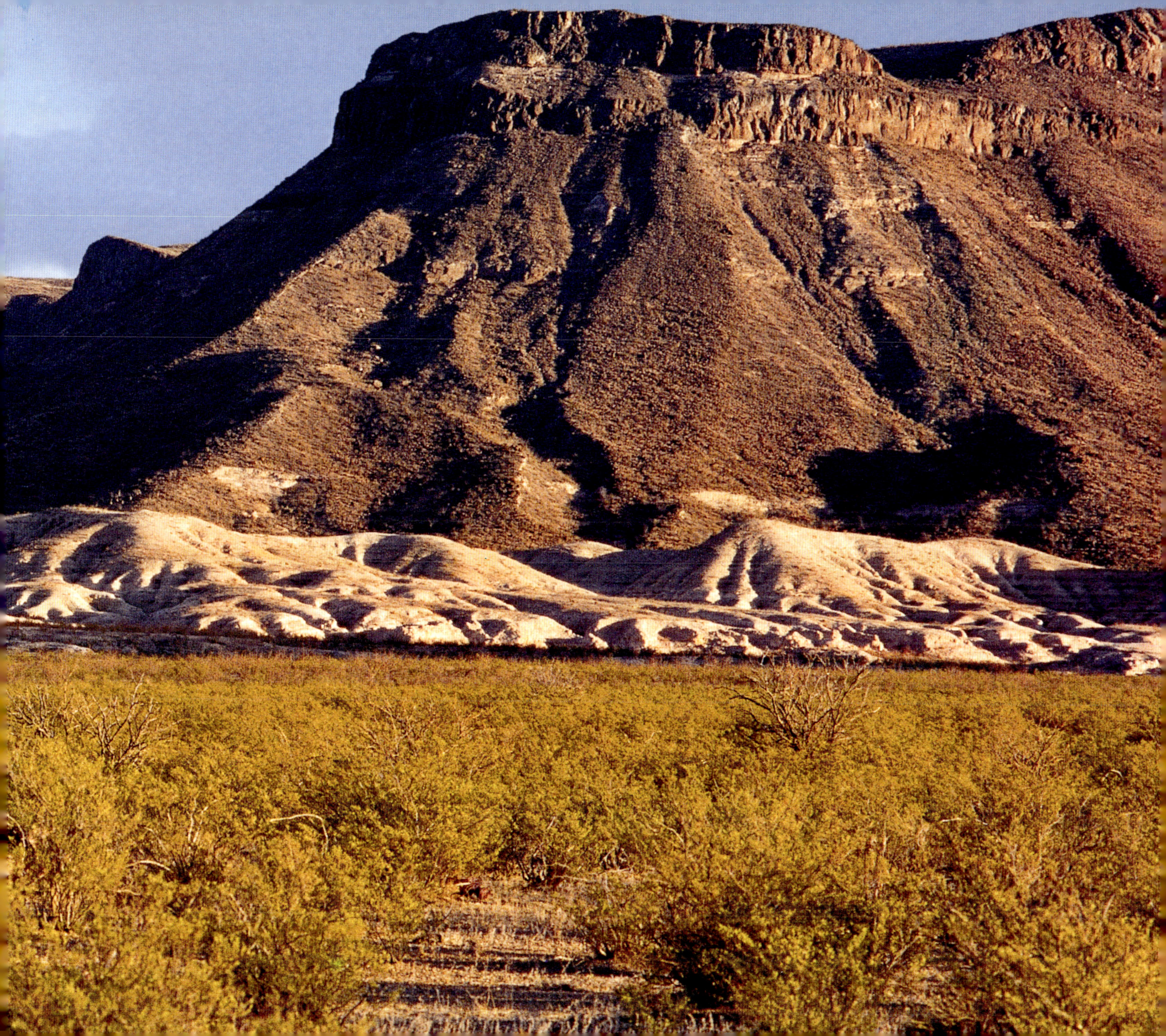

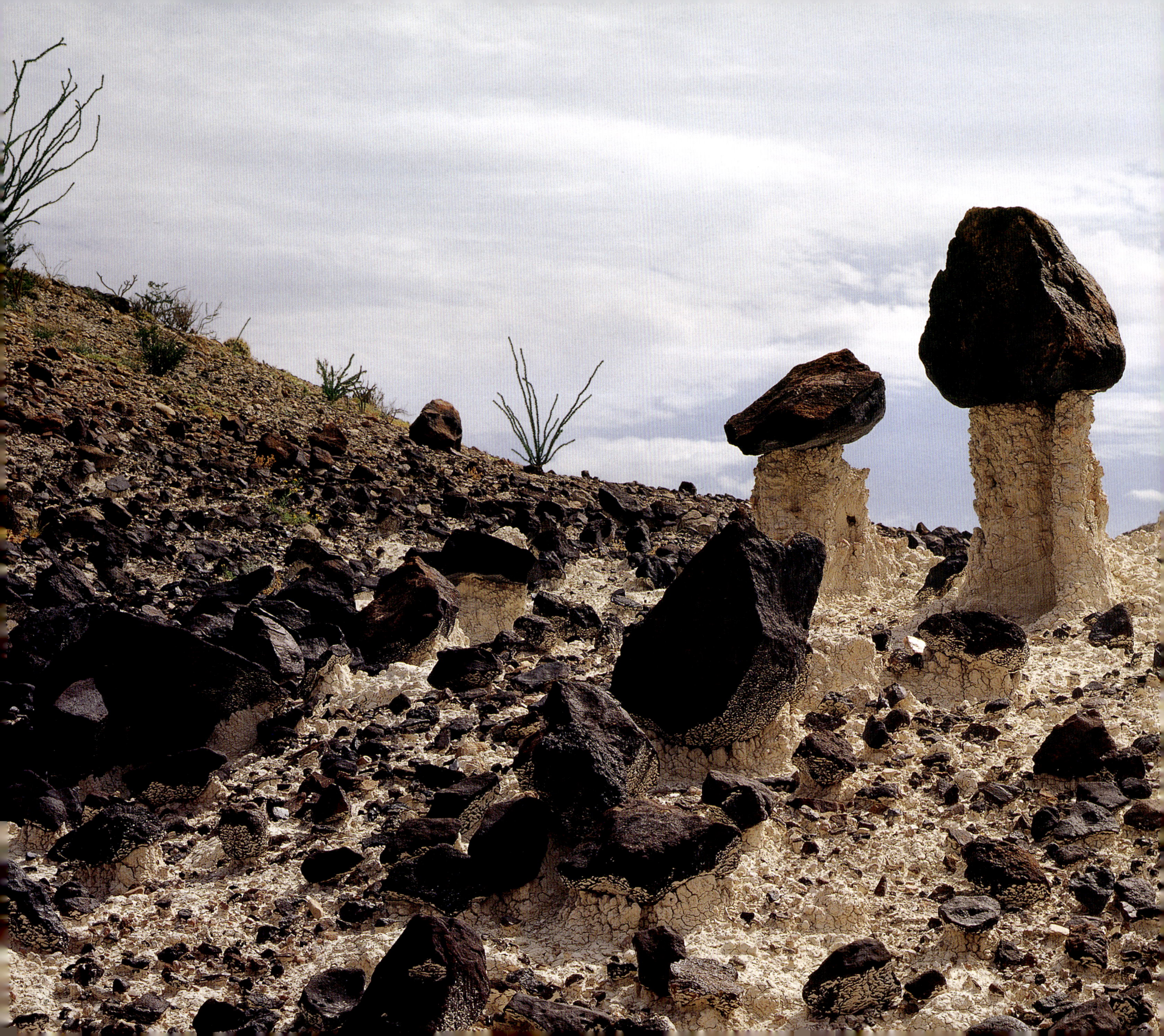

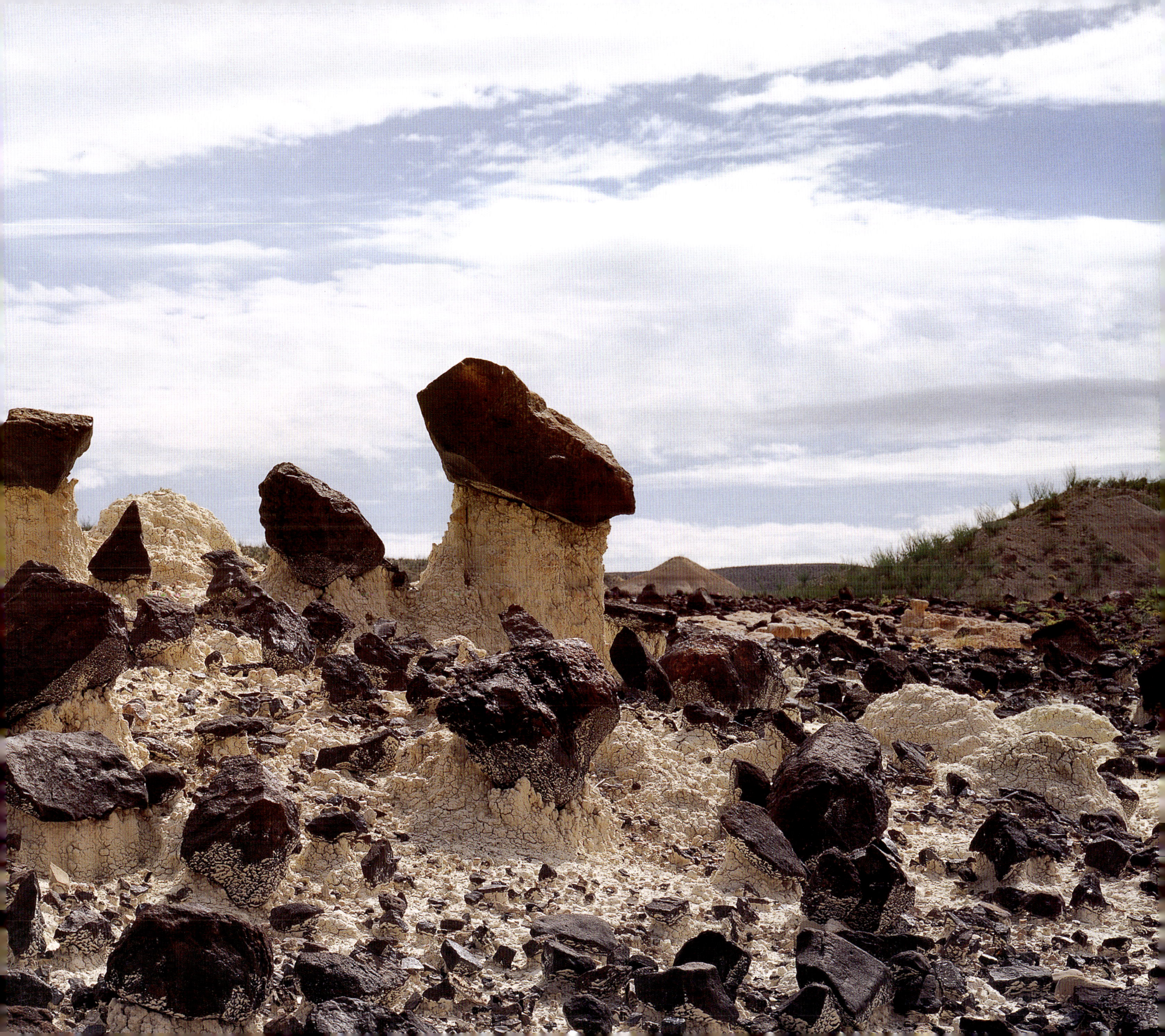

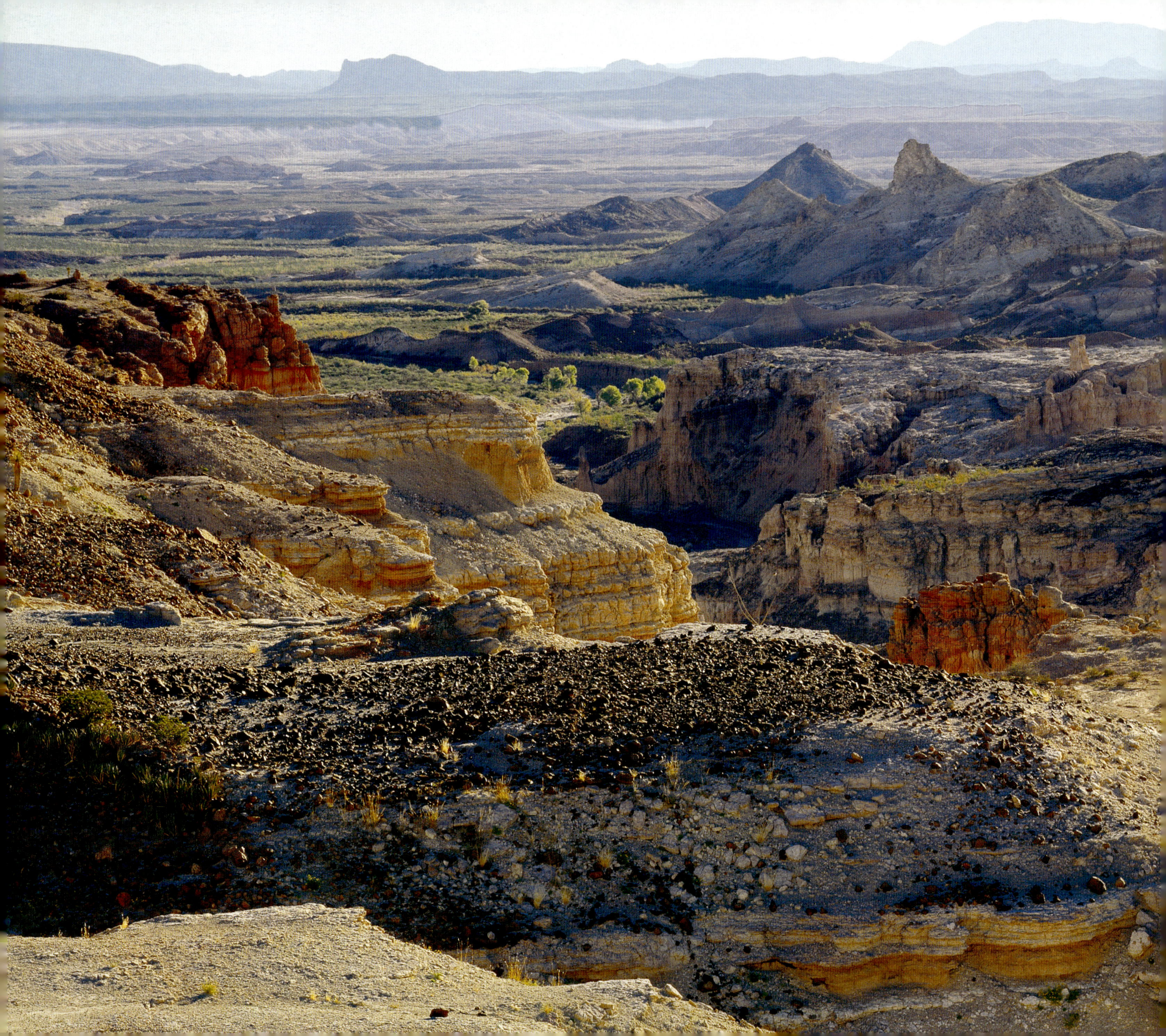

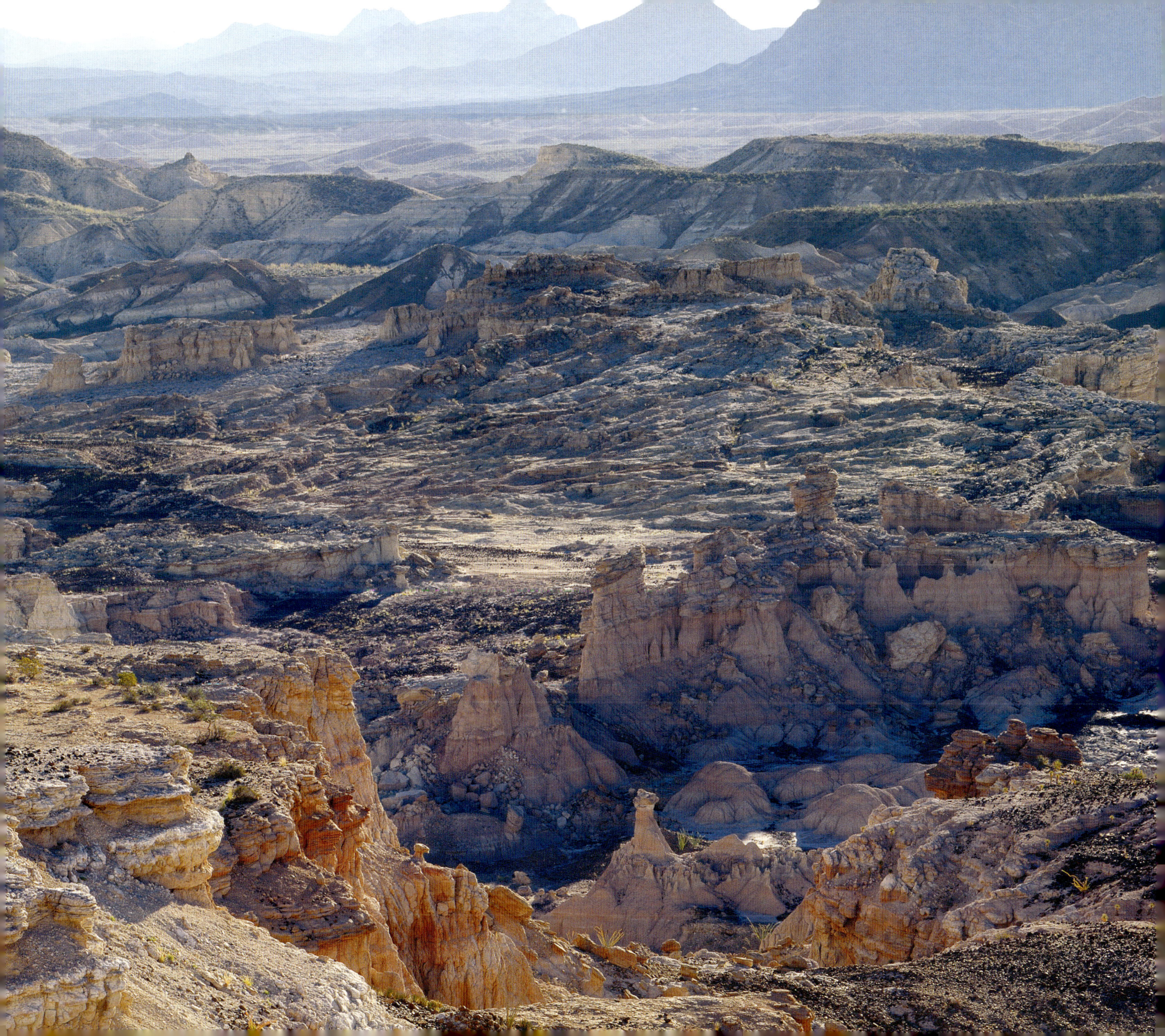

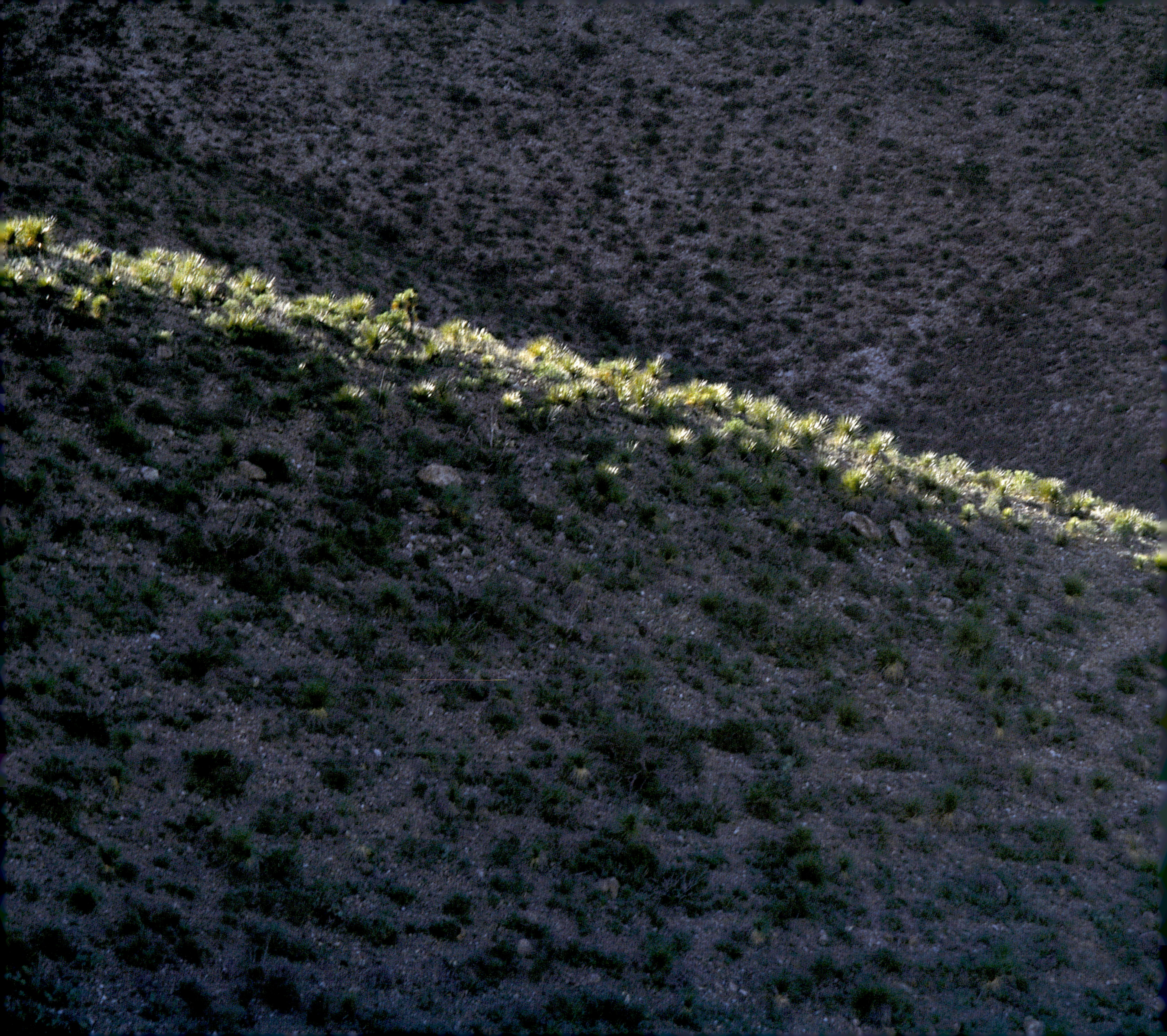

Ribbons of Time

THE DALQUEST RESEARCH SITE

PHOTOGRAPHS BY WALTER W. NELSON

TEXT BY DOUGLAS PRESTON

MIDWESTERN STATE UNIVERSITY PRESS

WICHITA FALLS, TEXAS

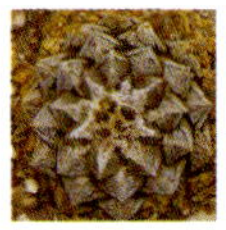

BOOK DESIGN BY MARY SWEITZER

ISBN 0-915323-15-X

FIRST PUBLISHED IN 2006 BY

MIDWESTERN STATE UNIVERSITY PRESS

OFFICE OF THE PRESIDENT

MIDWESTERN STATE UNIVERSITY

3410 TAFT BLVD.

WICHITA FALLS, TX 76308

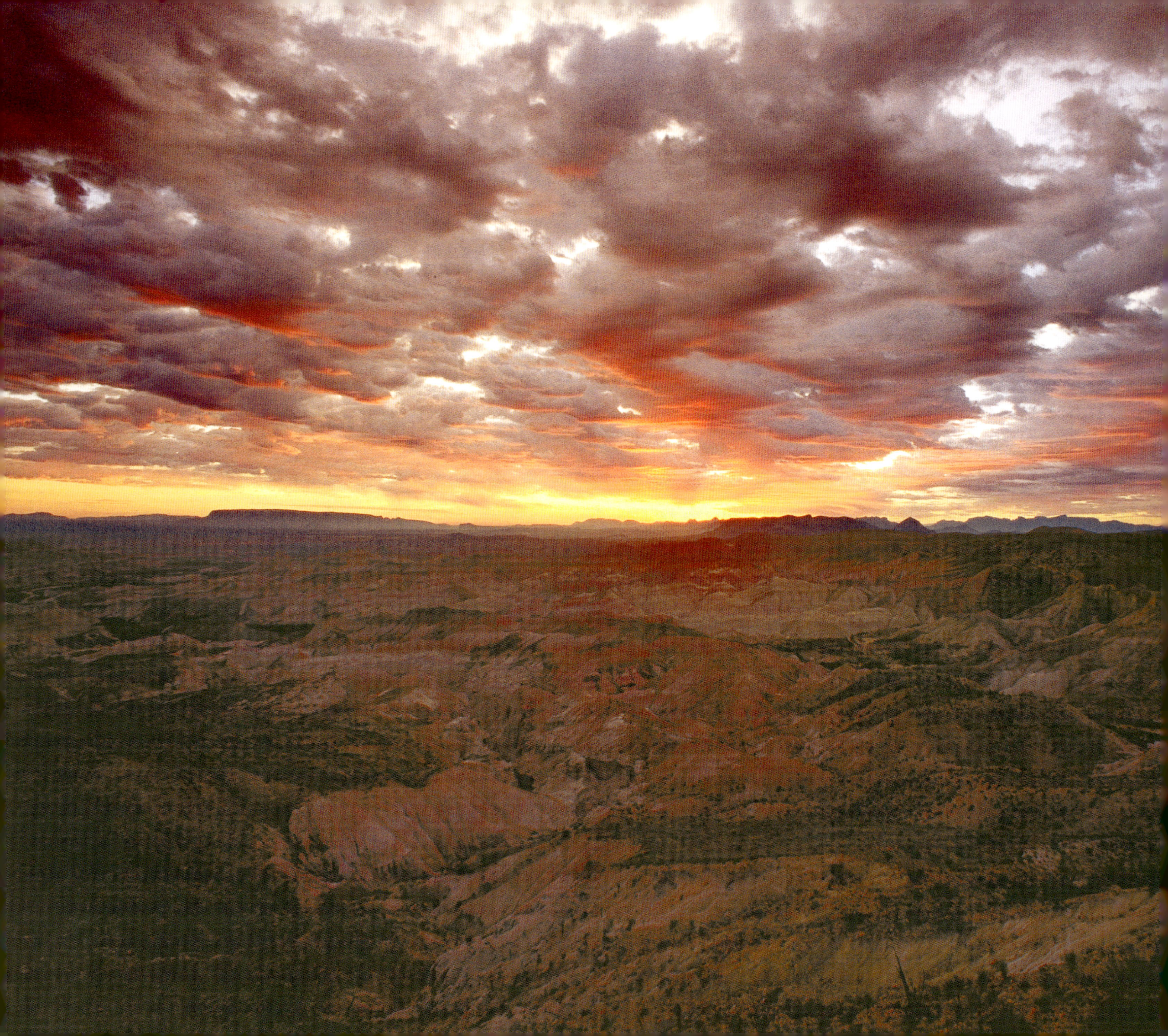

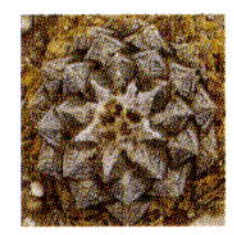

This book is dedicated to Dr. Jesse Rogers,
President of Midwestern State University

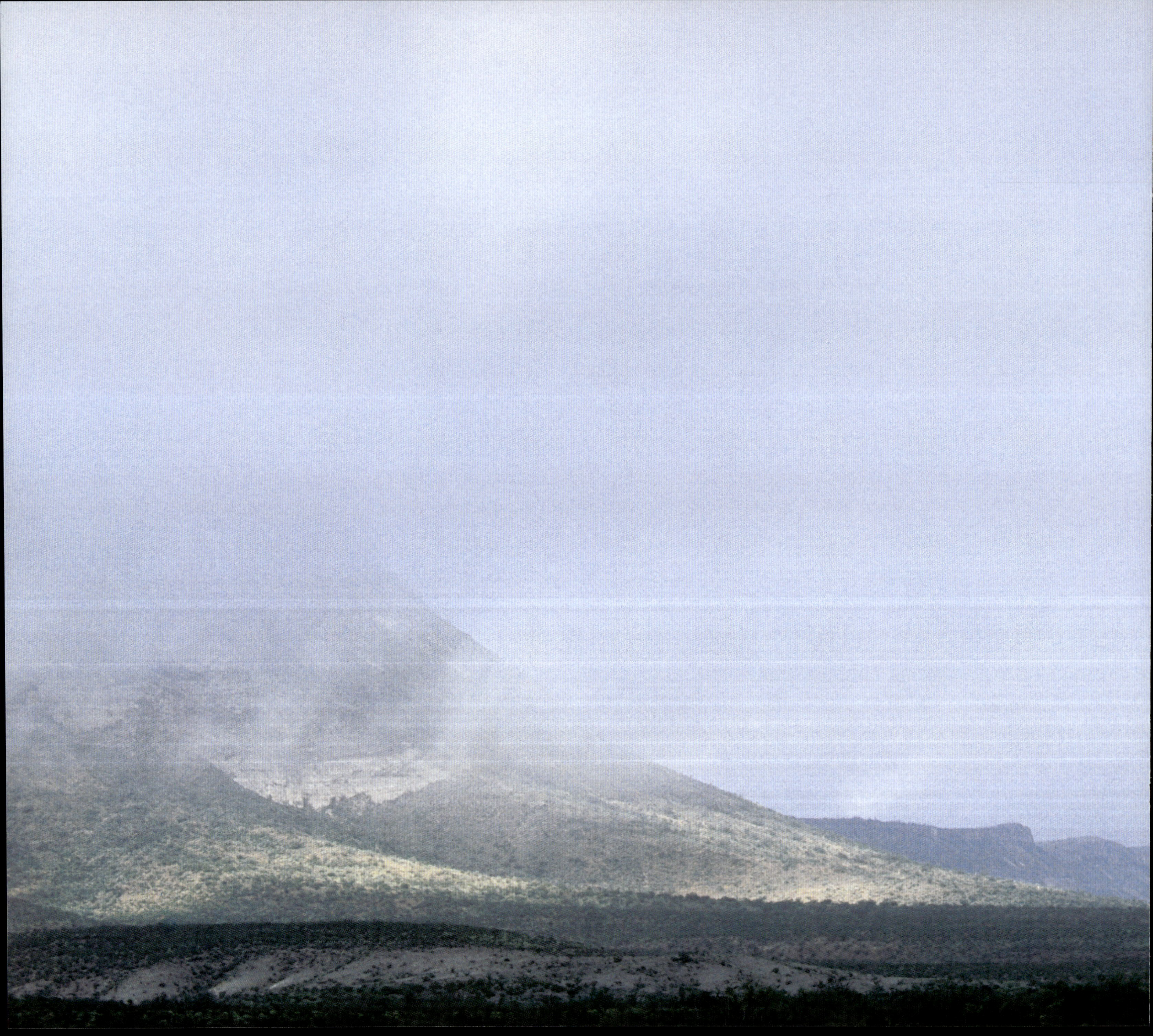

Foreword

JESSE W. ROGERS, PRESIDENT, MIDWESTERN STATE UNIVERSITY
NORMAN V. HORNER, DEAN, COLLEGE OF SCIENCE AND MATHEMATICS

A UNIQUE PIECE OF THE CHIHUAHUAN DESERT WAS DONATED TO MIDWESTERN State University in 1996 by the late Dr. Walter W. Dalquest and his wife Rose. The property, now known as The Dalquest Research Site, is used for scientific research. Rugged and pristine, it is a naturalist's dream. Rich in geologic formations and varieties of plant life, the area is daunting. Even a glance at its canyons tells one why the place has had little human traffic.

We realize the tremendous value of this gift and want the property to be studied for years. Thinking about the site and our university's history, we find it appropriate to mention the many accomplishments of Dr. Dalquest, one of our nation's great explorers and naturalists. His home was the field and his appreciation of the natural world was immeasurable. As a Professor of Biology at MSU, he taught hundreds of students and served as mentor to undergraduates, graduate students, and young faculty alike. During his career, seven of his books were published, along with 173 other research publications in the fields of mammalogy, ornithology, ichthyology, and vertebrate paleontology. He was also one of the world's authorities on Pleistocene fossil horses. His wife Rose supported his research in spirit and close companionship. We greatly appreciate her generosity in assisting with this publication and for numerous other gifts that benefit our university.

When the idea of publishing a work to describe the property began germinating, we knew we wanted a book that would provide the historical record of the land; we wanted the work to be educational as well as aesthetically effective. Many individuals, of course, contributed time and effort to help develop this book. Mrs. Lee Bennett, of Marfa, provided considerable background information and numerous contacts for Walter Nelson, the photographer, and Douglas Preston, the author of the narrative. Our colleagues Drs. William Cook and Pamela Stephens provided valuable identifications of plants and geologic formations. James Hoggard, Editor and Director of the MSU Press, has done his typical first-class job. His attention to detail, writing skills, and knowledge have helped us all.

Walter Nelson has done a remarkable job of capturing the natural beauty of the plants and the powerful imagery of the geologic formations on the property, often under harsh climatic conditions. Through an entire year he took numerous photographs to illustrate seasonal changes and variations of imagery caused by dramatic shifts of light. He spent more than 100 days camping in the canyons to find the best lighting possible to preserve in his photographs the tremendous upheavals and fanning slopes of a geologic microcosm. Douglas Preston, a well-known author of both fiction and nonfiction, has traced, with precision and gracefulness of phrase, what this site has been like for the last 700 million years.

The work presented here, in both word and image, is impressive. We have all learned important things from it, and we are confident that you will enjoy it as well.

Previous Pages: Devil's Graveyard formation, viewed from Dalquest peak, looking northeast.
Tascotal Mesa, northwest of site.
Totoms.
Devil's Graveyard, viewed from the white rim, looking southeast.
Early morning fog on the high plateau above the canyon.
Rim light from Dalquest peak in late afternoon, looking west.
Ribbons of water in motion.
Sunrise across the canyon from Dalquest peak, looking east.
Opposite page: Tascotal Mesa in fog, west of site.

Acknowledgments

BY WALTER W. NELSON

Above all I would like to thank Rose W. Dalquest of the Dalquest Foundation and Martha G. Fain of the Fain Foundation for their unfaltering belief in the project and for making this dream become a reality through their kindness, encouragement, and generous financial support. I thank Douglas Preston, my old traveling companion in saddle and on foot, for his colorful and artistic verbal expression, and David H. Gibson, close friend and photographer, with whom I have traveled for over eighteen years photographing in some of the most majestic places in the world—this being one of them—and for his belief in my work.

I am most grateful to Mrs. Lee Bennett of the Bandera Ranch for her strong compassion for the land and its history; Red and Waynelle Strachan of the Alazan Ranch for their horses, food, lodging, and knowledge of this rugged land; Kevin Giraud, cowboy and wrangler at the Alazan Ranch, for his assistance in packing in and out of the canyon; Linda Walker of the Lajitas Stables for her excellent pack mules Lucie and Maud, and for her unending supply of energy getting them to the canyon.

I would like to thank Peggy Jones and the late Pierre Stine for their assistance in making this adventure a little easier; Dr. Jesse Rogers, President of Midwestern State University, for believing in this dream; Dr. Norman Horner, Dean of the College of Science and Mathematics at MSU, friend of Walter W. Dalquest, and wise custodian of this magical land.

I am very grateful to Mary Sweitzer for her splendid book design; Richard Ash, former chairman of the Art Department, MSU, for his unfaltering belief in the project; James Hoggard, Director and Editor of MSU Press, for his valuable help; and I thank Joe N. Prothro, David White, B.C. Bennett, Evelyn Luciani, Jenny Gattis, Cindy Ashlock and all those others whose paths I crossed while making this dream become a reality.

I would like to express my gratitude and love to my mother, Dorothy W. Nelson, for allowing me the opportunity to travel an artistic journey in my life.

BY DOUGLAS PRESTON

I would like to add my expressions of gratitude to all the people Walter thanked above, in particular Dr. Norman Horner, Dr. Jesse Rogers, and Lee Bennett. I also would like to offer my great thanks to Walter himself for inviting me to participate in this splendid adventure, for showing me one of the most remarkable canyons in the world, and for being my friend, artistic teacher, and traveling partner these past two decades. I thank Dr. Pamela Stephens of MSU for her research into the geology of the canyon; Dr. William Cook, Chairman of the MSU Biology Department; Dr. Cecilia Thompson for her historical research; and Joy Rawls for her fascinating stories. And, as always, I offer all my gratitude and love to my wife Christine and my three children, Selene, Aletheia, and Isaac.

Opposite page: High plateau in fog.

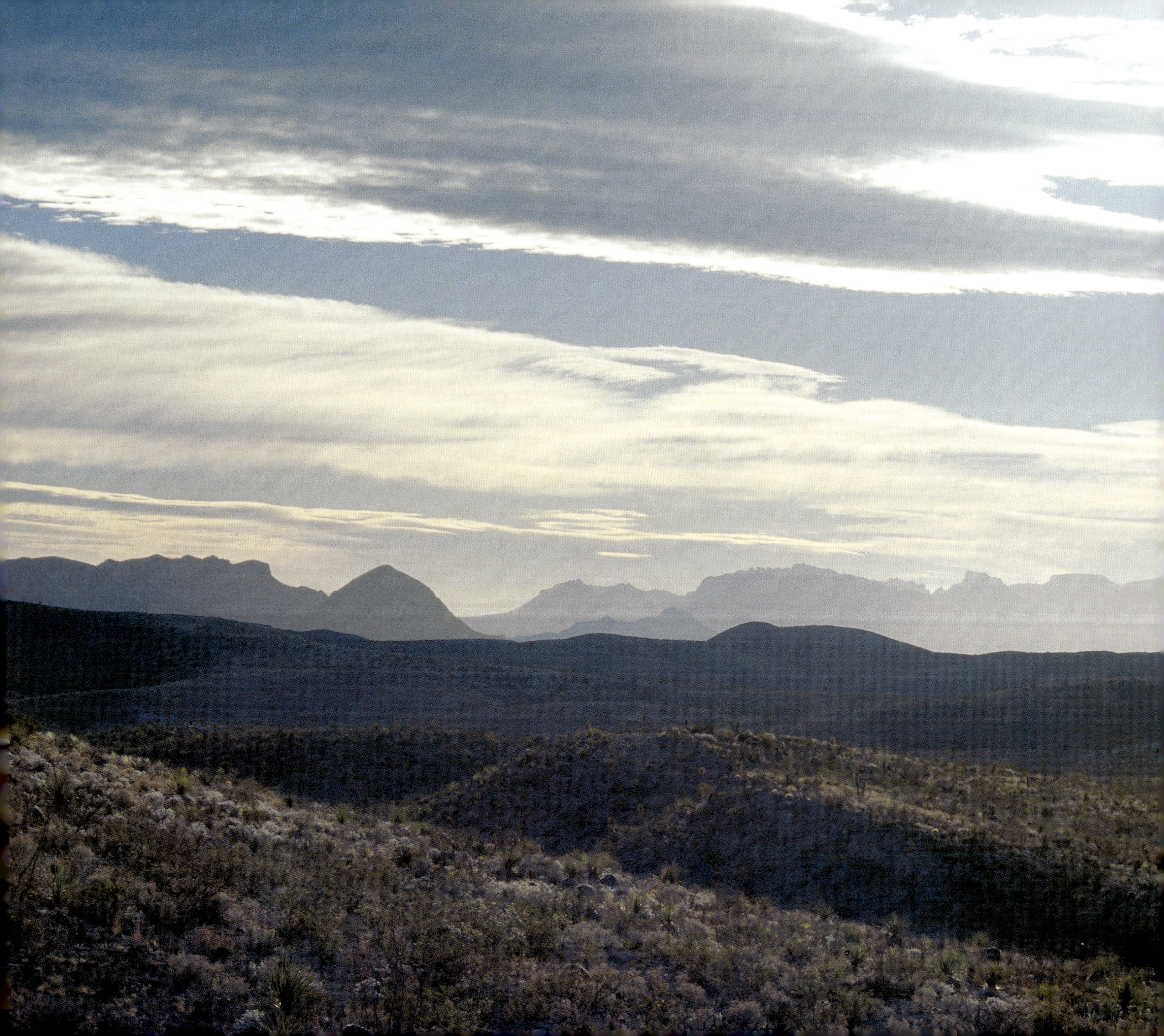

The House Made of Time

BY DOUGLAS PRESTON

IF YOU TRAVEL DUE SOUTH FROM MARFA, TEXAS, ALONG ALAMITO CREEK, YOU pass through typical Trans-Pecos Texas ranching country—vast undulating blankets of grass thrown down under blue skies, bordered by low hills and thick with mesquite. Thirty miles south of Marfa, the terrain begins to change dramatically. The Rio Grande is still more than sixty miles away, but already the land breaks gently southward toward the river. The asphalt road turns to dirt and the grass gives way to creosote bush and cholla. Strange black mountains begin jutting over the horizon—the eroded throats of dead volcanoes. At each dip in the road you see where flash floods have ripped across, cutting gullies and leaving behind sand and rocks.

As the road gets rougher, the creosote bushes are joined by other plants typical of the Chihuahuan Desert—lechuguilla, ocotillo, prickly pear, and yucca. You are now deep in the Big Bend of Texas, and here you have the strong sense of leaving civilization behind. This is a piece of primeval America, a landscape little changed for ten thousand years.

As you continue south, each ridge becomes steeper than the last, and at each crest you glimpse a distant sea of jagged mountains, purple in the shimmering heat. Soon you are surrounded by sheer-walled mesas and volcanic plugs, some standing more than a thousand feet above the plains, sculpted by rain and wind into malefic shapes. The road deteriorates until it is little more than a set of tracks strewn with axle-breaking boulders.

You reach the last ridge, and you can see where the track curves back on itself to form a turnaround. You are now sixty-five miles from Marfa, Texas. Cell phone coverage ceased 40 miles back. There are no trees or water. Where the track ends, the land disappears and there is nothing but blue sky. You feel as if you have reached the edge of the world. You step out of the car into a silent landscape. The friendly sigh of wind in trees or rustling through grass cannot be heard, only the itchy rattle of creosote. The light is so intense that the landscape seems as bleached and hollow as a cowskull. It is just a few steps to the edge, and then you will see what lies below.

The cowboys who worked cattle above this rim called it the Falls, because the land dropped away as suddenly as a waterfall, and because hundreds of real waterfalls poured off the cliffs into the canyons below after a heavy rain. The cowboys below the rim called it the Devil's Graveyard, and the strange rock formations the Devil's Fingers, because from their vantage point it looked like Hell itself. In 1996, this landscape acquired a new name: The Dalquest Research Site. In that year it was given to Midwestern State University by Walter and Rose Dalquest to be preserved in perpetuity as a place for scientific research. Dalquest, a long-time professor at MSU and a nationally known mammalogist and paleontologist, realized that the site, lying in the heart of the Big Bend of Texas, would make a unique laboratory for scientific study due to its spectacular geology, its unspoiled environment, and its rich panoply of rare Chihuahuan Desert plant and animal life.

Opposite page: Midmorning light, looking southeast, from Dalquest peak.

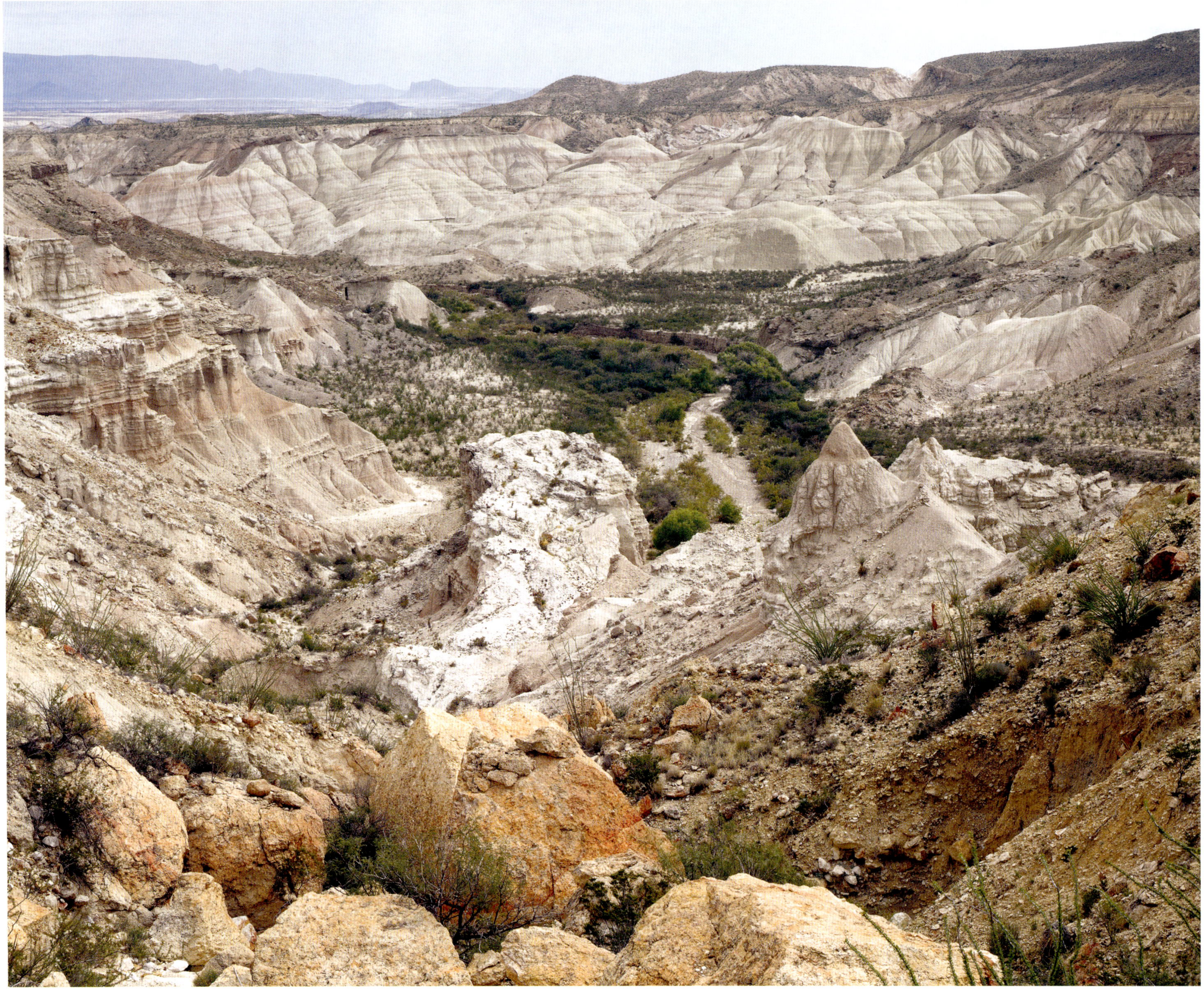

I first made this journey to the Dalquest Site in September, 2003, in company with Walter Nelson, the photographer of this volume. Nelson had spent a year meticulously documenting the area in 35mm, 4x5 and 8x10 large-format photography. He knew the site well; he had built a pack trail down the treacherous cliffs into the area and established two base camps. Like Ansel Adams, he had packed his large-format cameras to the bottom on muleback and had spent more than a year, off and on, in the canyon photographing across the seasons. Walter was there to photograph yet again; I was there to observe.

The first time you stand on this rim and look down, what you see defies comprehension. Three canyons merge into one, creating a massive train-wreck of rock, a jumbled chaos of stone—labyrinthine, alien, slashed open by erosion and bleeding every color of the rainbow. The eye roams over a wilderness of bizarre rock formations—spires, knife-edges, banded cliffs hanging like draperies, pillars, skull-like hills, snow-white flats, balancing rocks, arches and needles. At first you have no sense of scale—it could be as big as the Grand Canyon. It is only after a few minutes that you realize you are gazing at an area of no more than two square miles, into which has been packed every imaginable geological shape, form, color, and texture.

War and Peace

IN THE EARLY 1950s, A GEOLOGIST WITH THE PHILLIPS PETROLEUM COMPANY came through this country, armed with a scintillation counter, prospecting for uranium. He climbed the highest point of land around, sat down on a rock, and simply stared. After five minutes, the rancher who was with him wondered why he was so absorbed.

"What do you see?" he asked.

"An unbelievable story," the prospector murmured. "A fascinating story."

To those who can read the rocks, the Big Bend landscape recounts a tale of epic proportions, a geological *War and Peace* that spans more than seven hundred million years. It is a story of plains thundering with herds of dinosaurs, of lush swamps and primeval forests, of shallow seas teeming with strange life forms, of massive volcanic eruptions, earthquakes, floods, and lava flows. So varied is the geology of the Big Bend that it contains every known kind of mountain in the world. If the geological history of the Big Bend is *War and Peace*, then the Dalquest Site would have to be the battle scene—the most violent chapter in the novel.

Forty-nine million years ago, before the cataclysm that created the Dalquest Site, we would have found an Eden here, a gentle land watered by meandering rivers and oxbow lakes, overspread with cool forests of sycamores, cedars, and giant tree ferns. Open areas would have been covered with flowering plants, as grass was just then beginning to evolve from an ancient type of lily. The rivers and lakes teemed with garfish, turtles, and a species of giant crocodile with jaws four feet long. The climate was idyllic, with cool wet summers and mild winters. In the distance stood a line of jagged blue peaks wreathed in snow—a southern spur of the freshly uplifted Rocky Mountains.

Opposite page: View of the trail, looking east into the South Fork of Alamo de Cesario Creek.

The dinosaurs had become extinct sixteen million years before. The ecological niches once dominated by these great reptiles were now being filled by an array of bizarre mammals that looked like nothing alive today. They included the ancestor of the horse (about the size of a terrier), a hippo-like animal that spent its days wallowing in the mud, and the strange phenacodus, which looked like a cross between a tapir and a bear, if such a creature can be imagined.

It would have been a peaceful scene, disturbed only by the presence of a few volcanoes, smoking lazily under blue skies. But the seeds of destruction were growing deep underground, where two plates of the earth's crust were in slow, agonizing collision. The Pacific Plate was being shoved underneath the North American Plate, and the pressure and movement were melting vast amounts of rock, forming giant underground pools of magma. Over time the magma, under enormous pressure, wormed its way upward through fractures in the rock.

One fateful day about forty-nine million years ago, a tongue of magma reached the surface near present-day Paisano Pass, about sixty miles north of the Dalquest Site, erupting with explosive force. It ejected millions of tons of ash, at times casting such a pall over the landscape that noon was as dark as midnight. The ash blanketed the land, burning and burying the forests and killing the animals. Overnight, Eden turned into a lifeless desert.

For seventeen million years volcanoes at Paisano Pass and elsewhere erupted at regular intervals, sometimes with incredible violence, spewing ash as far as present-day Georgia. Rivers and occasional floods pouring out of the nascent Rocky Mountains and Mexico cut channels through the ash, leaving behind beds of limestone cobbles, sand and massive volcaniclastic deposits—layers of volcanic ash swept up by water and redeposited. Shallow freshwater lakes formed, only to be suffocated and buried by the next eruption. The bones of dead animals were sometimes caught up in periodic floods, to be deposited as fossils embedded in these beds of ash.

Countless eruptions and floods laid down hundreds of layers. Then dikes of denser magma forced their way through cracks to the surface, gushing out in lava flows which spread in incandescent rivers across the land, blasting up huge eruptions of steam where they encountered lakes and streams, and hardening into black basalt. At least nine successive lava flows capped the Dalquest Site.

Seventeen million years after it began, the geological violence came to an end. The original landscape had been buried under 1500 feet of ash and hundreds more of lava. Over time the soft beds of ash hardened into rock. Each layer of ash contained slightly different trace elements, which tinted each layer a different color—from greens and blues to oranges, pink, reds, yellows, purples, whites, and browns—and which would later give the area its spectacular coloring.

When the volcanism ceased, life returned to the landscape. But here the Dalquest chapter in the geological tale of the Big Bend comes to an end. Layers laid down on top were later stripped away by erosion, their stories lost. For thirty-two million years, the land lay buried, hidden, waiting to be opened like a block of marble awaiting the chisel of a sculptor. Nature's chisel—water—was ready.

Opposite page: Forty million year-old sedimentary deposits east of trail.

The Rio Grande, twenty-five miles to the south, had already begun to slice a broad arc through the mountains, forming the Big Bend. A web of streams draining from the north fed the river. One of these feeder streams

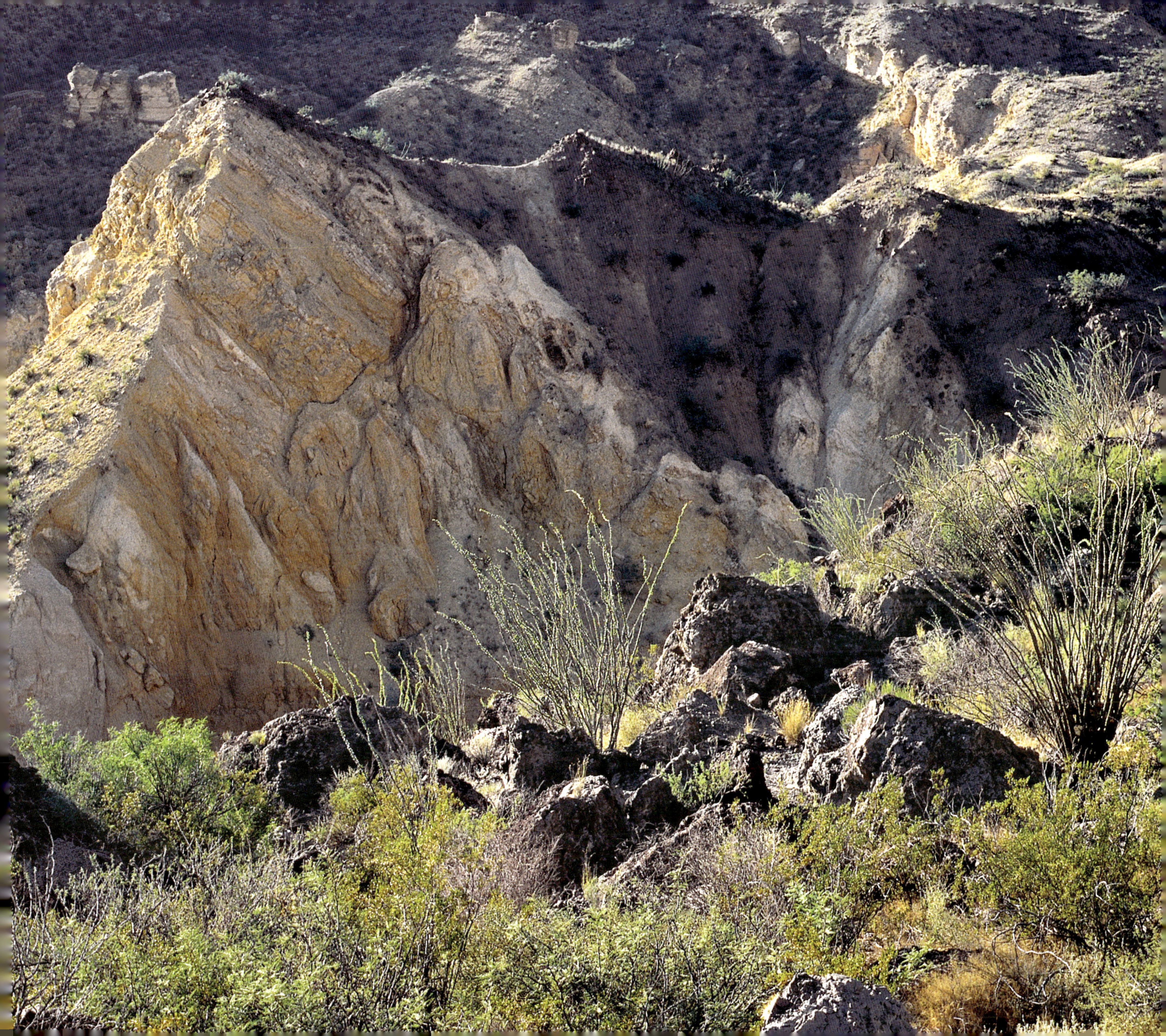

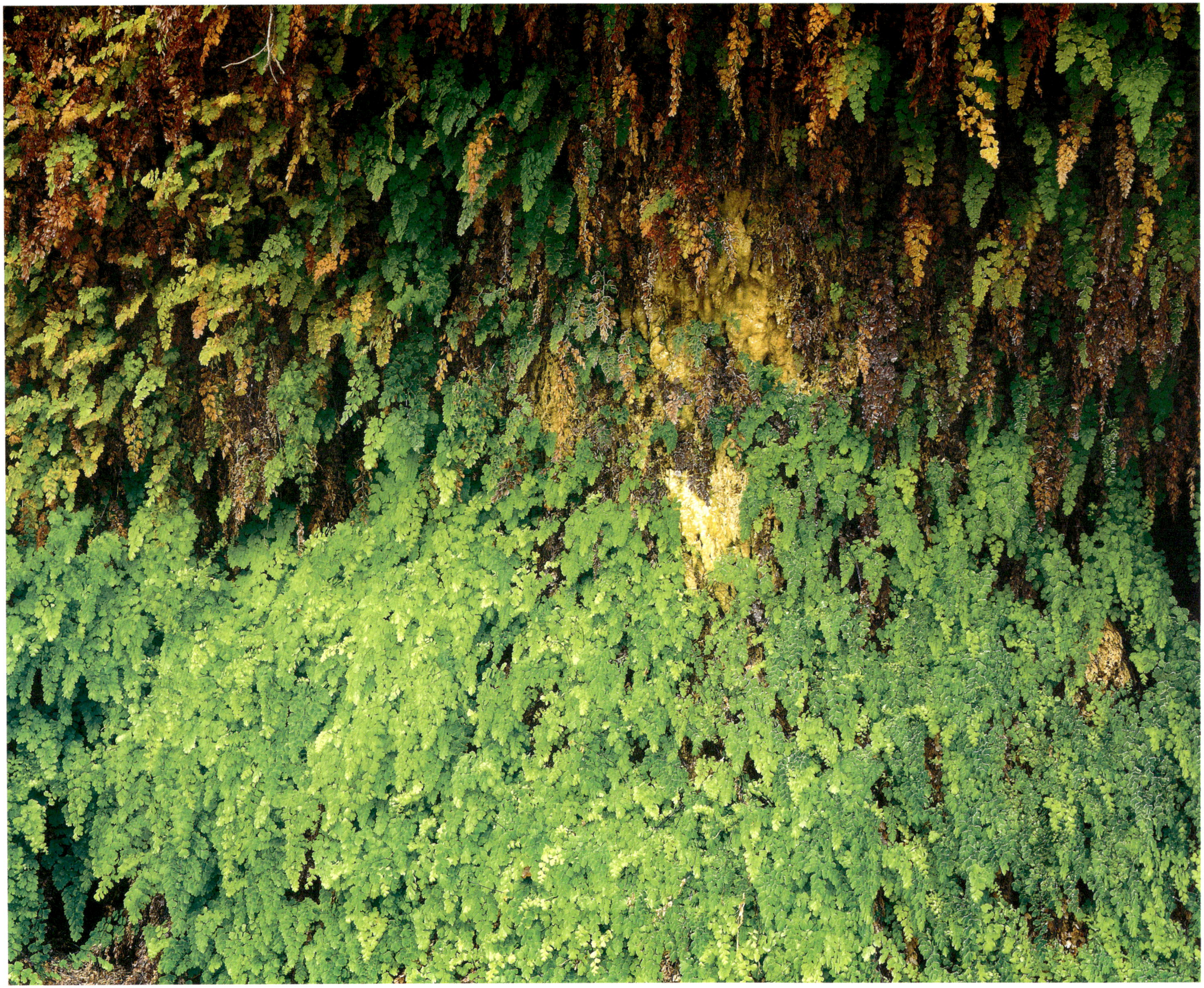

was the ancestor of Alamo de Cesario Creek, which cuts southeast across the Dalquest Site on its way to the Rio Grande. The creek probably ran year-round, meandering through rich grassy plains and forests inhabited by a familiar range of Ice Age mammals—saber-tooth cats, horses, camels, giant ground sloths, and mammoths.

Over millions of years, Alamo de Cesario Creek began cutting into the land downstream of the Dalquest Site, the erosion gradually working its way headward. It took a while for the slow action of the water to wear through the layer of basalt capping the site. But when the water hit the many layers of volcaniclastic ash underneath the basalt, it ripped through them like a hot knife through butter, sculpting the striped rock into curious arches, fins, blades and pillars, creating the geological wonderland we find today. Geologists, who are perhaps the most poetic of scientists, call these striped layers the Devil's Graveyard Formation.

Mammoths, Apaches, and Conquistadors

THE DALQUEST SITE FORMS A SMALL BUT SIGNIFICANT PART OF THE SPECTACULAR landscape of the Big Bend. Three waves of people settled this land: the Indians, the Spanish, and finally the Anglos. All three were awestruck by this landscape. The Apaches believed it was a lost corner of creation. The Mexicans called it the *despoblado*, the uninhabited place, and looked on it with bafflement and fear. One Mexican vaquero described the country in this way: "You go south from Fort Davis until you come to the place where the rainbows wait for rain, and the big river is kept in a stone box, and the water flows up hill. And the mountains float in the air, except at night, when they run off to play with other mountains." European and American travelers expressed similar feelings. One French traveler wrote in 1896: "Nowhere have I found such a wildly weird country. The very silence is oppressive. A man grows watchful for his own safety and becomes awe-struck by nature in her lofty moods." Another wrote, "Words cannot convey the strangeness and supreme melancholy of the landscape." Walter Prescott Webb, the man who led the drive to establish Big Bend as a national park, understood the landscape as well as any. In 1937, he wrote, "Perhaps order once prevailed here, but some mighty force wrecked the place, shook it down, blew it up, and set it afire ... Men of literature can not confine on the printed page the essential quality of the land, or convey the sense of unreality and romance that overwhelms the spectator."

Human beings first entered the Big Bend about thirteen thousand years ago. They were the Clovis people, perhaps the greatest hunters North America has ever seen. Their favored prey was the Columbian mammoth, the largest land animal our species ever hunted. This prehistoric elephant stood up to thirteen feet at the shoulder, weighed up to nine tons, and could run twenty-five to thirty miles an hour. Its much smaller descendant, the African elephant, is a dangerous animal to hunt even with a high-caliber rifle. The Clovis people used only spears and atlatls (a short spear launched at the end of a throwing stick) at very close quarters to kill mammoths. (The bow and arrow wouldn't be invented until thousands of years later.) It took a dozen or more spear thrusts to bring down a single animal—no mean feat when dealing with an enraged beast the size of a Mack truck. Cowboys working cattle still sometimes find Clovis spear points in the Big Bend area.

Opposite page: Water-weeping ferns growing on tufa at Fern Springs.

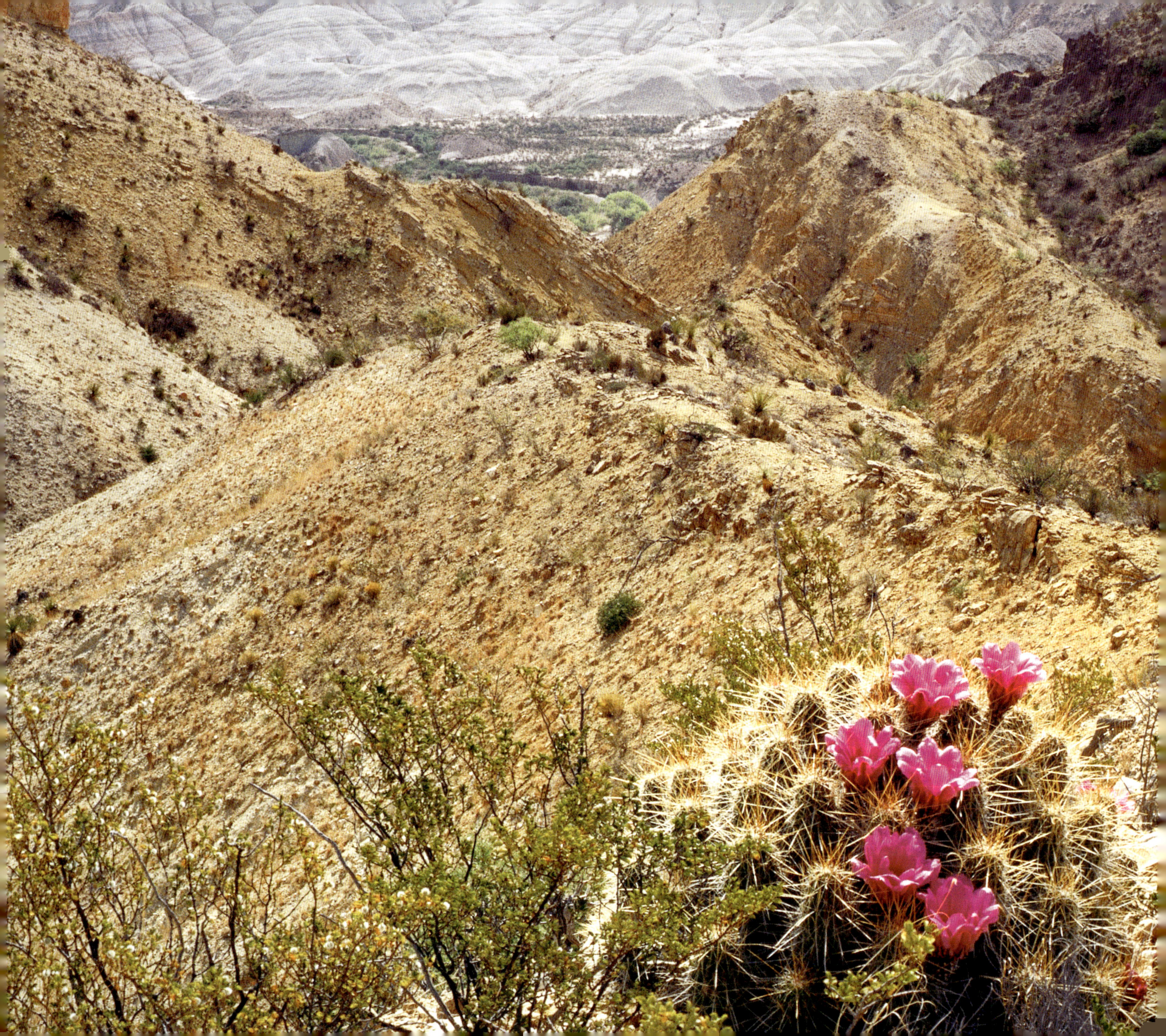

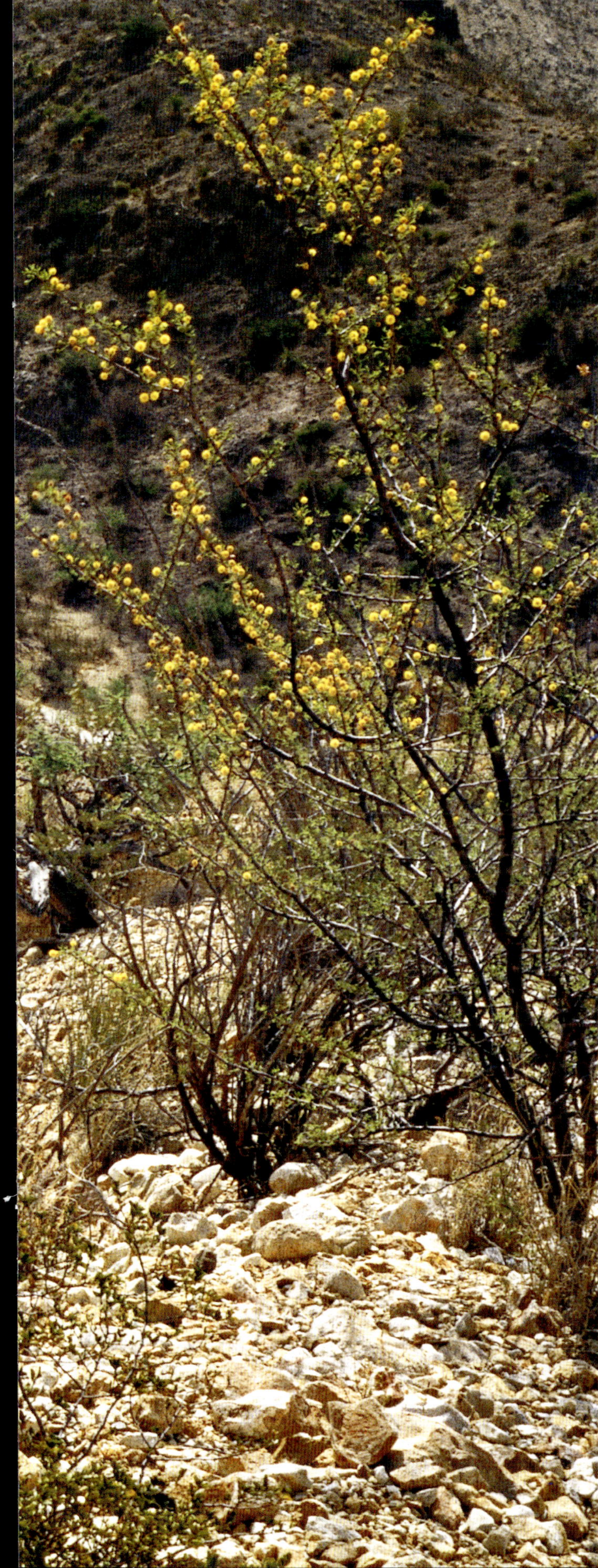

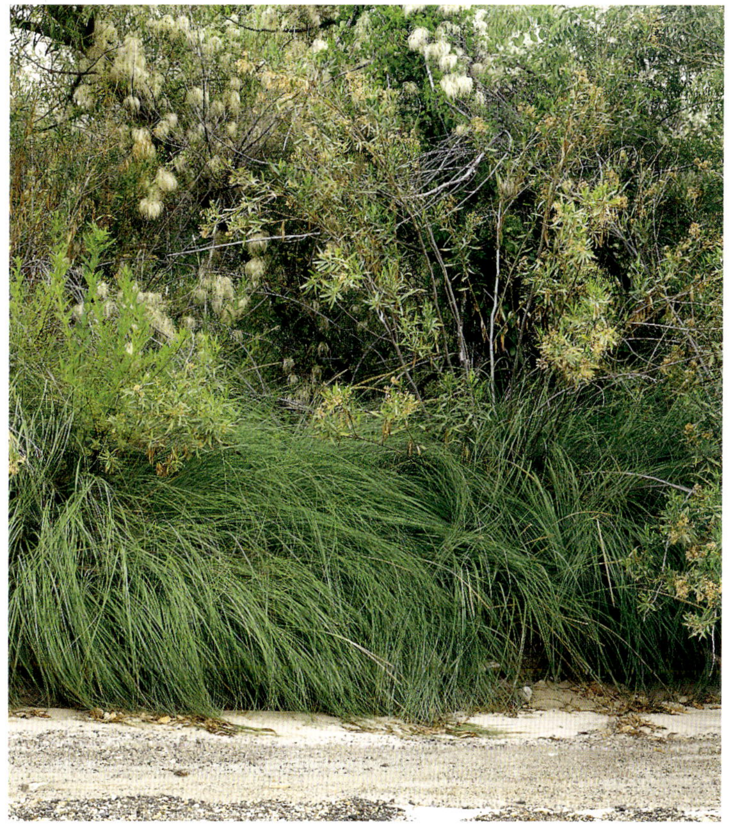

Opposite page: Trail into lower Devil's Graveyard.
This page: Desert vegetation along South Fork,
Alamo de Cesario Creek.

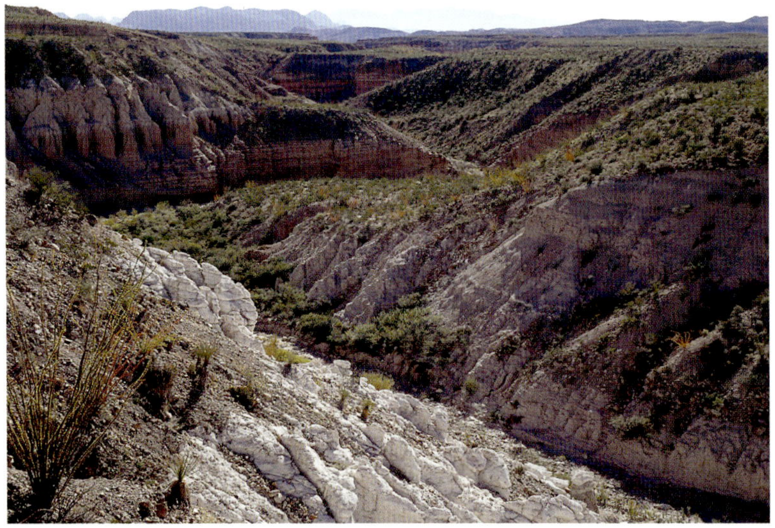

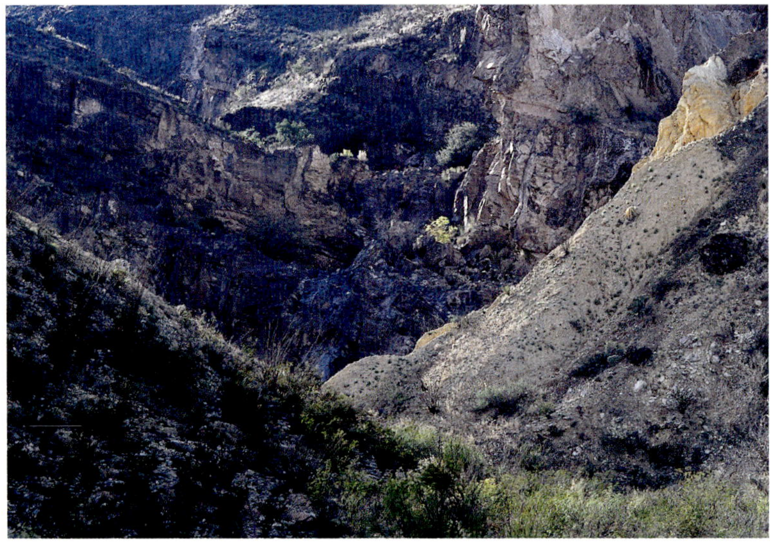

This page, above: Red canyon headwaters.
Below: Formation above Alamo Spring, looking west.

The Clovis people probably lived and hunted along Cibolo and Alamito Creeks west of the Dalquest Site; in those days they would have been year-round streams, dammed up by beavers and lined with groves of cottonwoods. Herds of elephants probably migrated along these north-south draws, spending the winter along the Rio Grande and then heading northward toward present-day Marfa during the summer.

Elephants beat down an excellent trail. Archaeologists believe that at least some of the famous historic trails in America started as mammoth trails. This may have been the case with the trails along Cibolo and Alamito Creeks, which would later become part of the Chihuahua Trail, one of the most important trade routes in Texas history. During prehistoric times Indians used this trail for trading salt, seeds, hides, macaw feathers, and slaves. Later this same trail would be used by Spanish conquistadors, American explorers, Apache raiding parties, and the railroad. It would be called by many names: the Salt Trail, Connelly's Trail, the Apache Trail, the Chihuahua Trail, the Old Fort Leaton Wagon Road, and finally the Panhandle and Santa Fe Railroad. This thirteen-thousand-year-old route of commerce and travel, passing less than twenty miles northwest of the Dalquest Site, would have a profound effect on its history.

The first Europeans arrived in the area in December of 1535, possibly on this very trail. The party consisted of three ragged Spaniards and a Moor, the only survivors of a six-hundred-man expedition to explore Florida. These four men were about to become the first Europeans to walk across North America—a stupendous feat unequalled in the annals of exploration. It would take them eight years. One of the men, Alvar Núñez Cabeza de Vaca, later wrote an account of their journey across this wild and unknown continent. He endured captivity, enslavement, starvation, and torture. The Indians at one point forced him to doctor one of their sick who had failed to respond to traditional medicine. Cabeza de Vaca murmured a few Christian prayers over the man and miraculously healed him. A great cry went up and the Indians hailed him as a medicine man. Word of his healing powers spread wildly, catapulting him from slavery to celebrity. Entire towns decamped to follow him as he traveled from village to village, healing with prayer.

In this manner, trailing a large Indian entourage, Cabeza de Vaca became the first European to enter the Big Bend. At the Rio Grande he encountered large villages of Jumanos Indians living in circular rock structures. Such was his fame that when he arrived he found the Indians seated in their houses, bowed with their faces turned toward the wall, their hair pulled over their eyes, and all their worldly goods piled up as an offering to him. After remaining a while in the Big Bend, Cabeza de Vaca pressed on, looking for his own people. When he finally reached Mexico City, his story of crossing such a wide expanse of America caused huge excitement, especially the rumors he had heard of immensely rich cities in the interior. These rumors triggered Coronado's famed search for the Seven Cities of Gold and the subsequent Spanish exploration of the American Southwest.

For the next three hundred and fifty years, the Big Bend area saw fitful exploration but little settlement, except along the course of the Rio Grande. The vast mountains and deserts of the interior, of which the Dalquest Site was part, were the terrifying *despoblado*, the desolate, empty land. But the word *despoblado* was something of a misnomer. Originally the land was inhabited by the Chisos Indians, who were forced out by the Mescalero

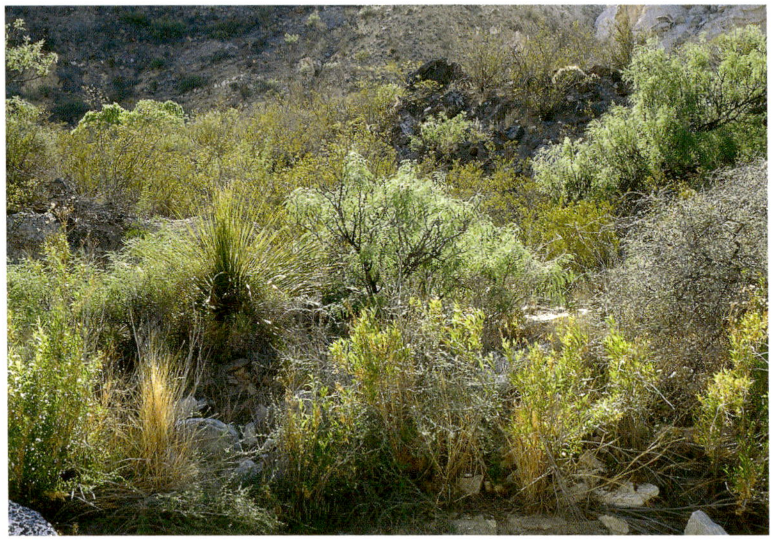

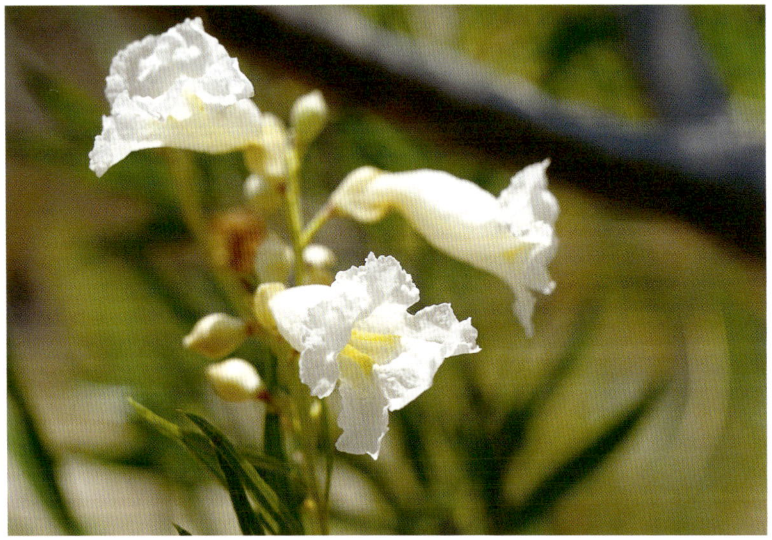

This page, above: Desert vegetation near upper spring.
Below: White desert willow in spring.
Opposite: Early morning, looking south
through cottonwoods into furrowed volcanic ash.

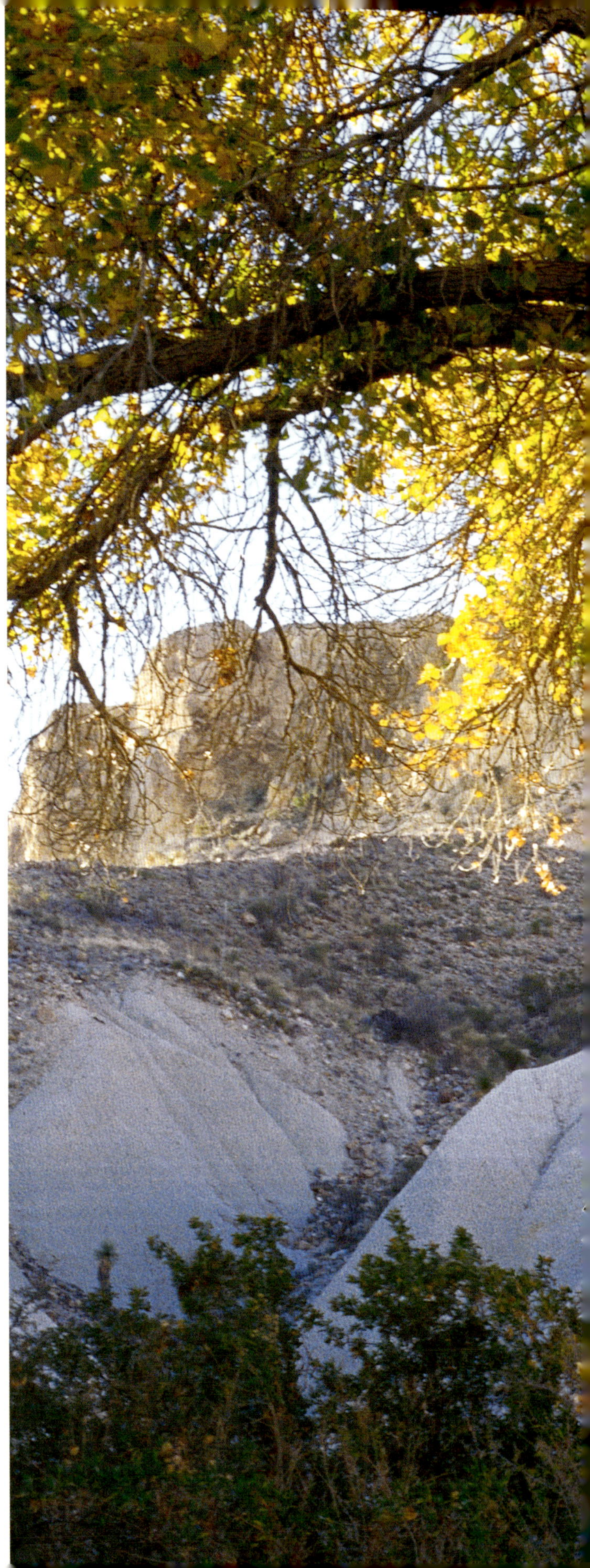

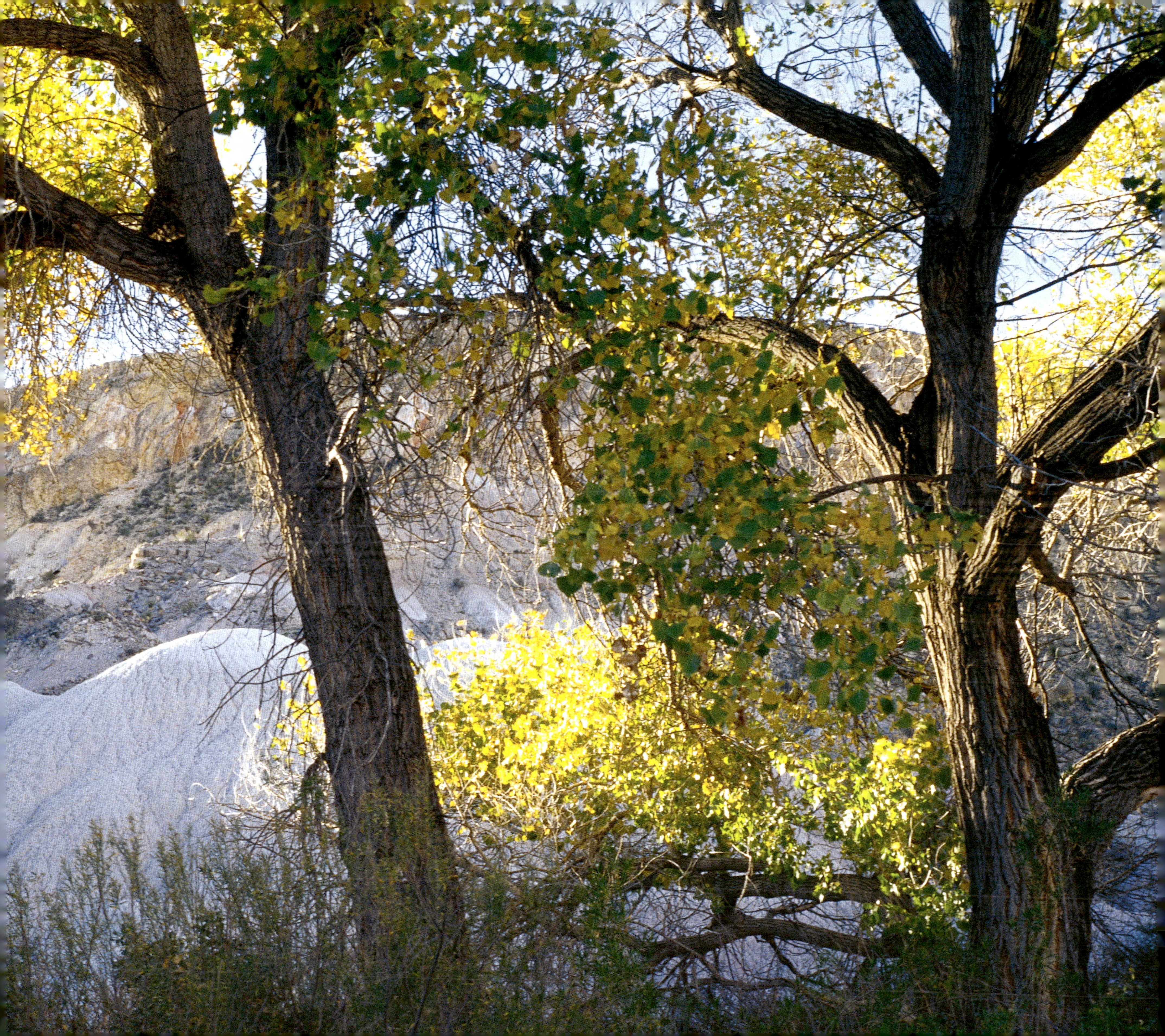

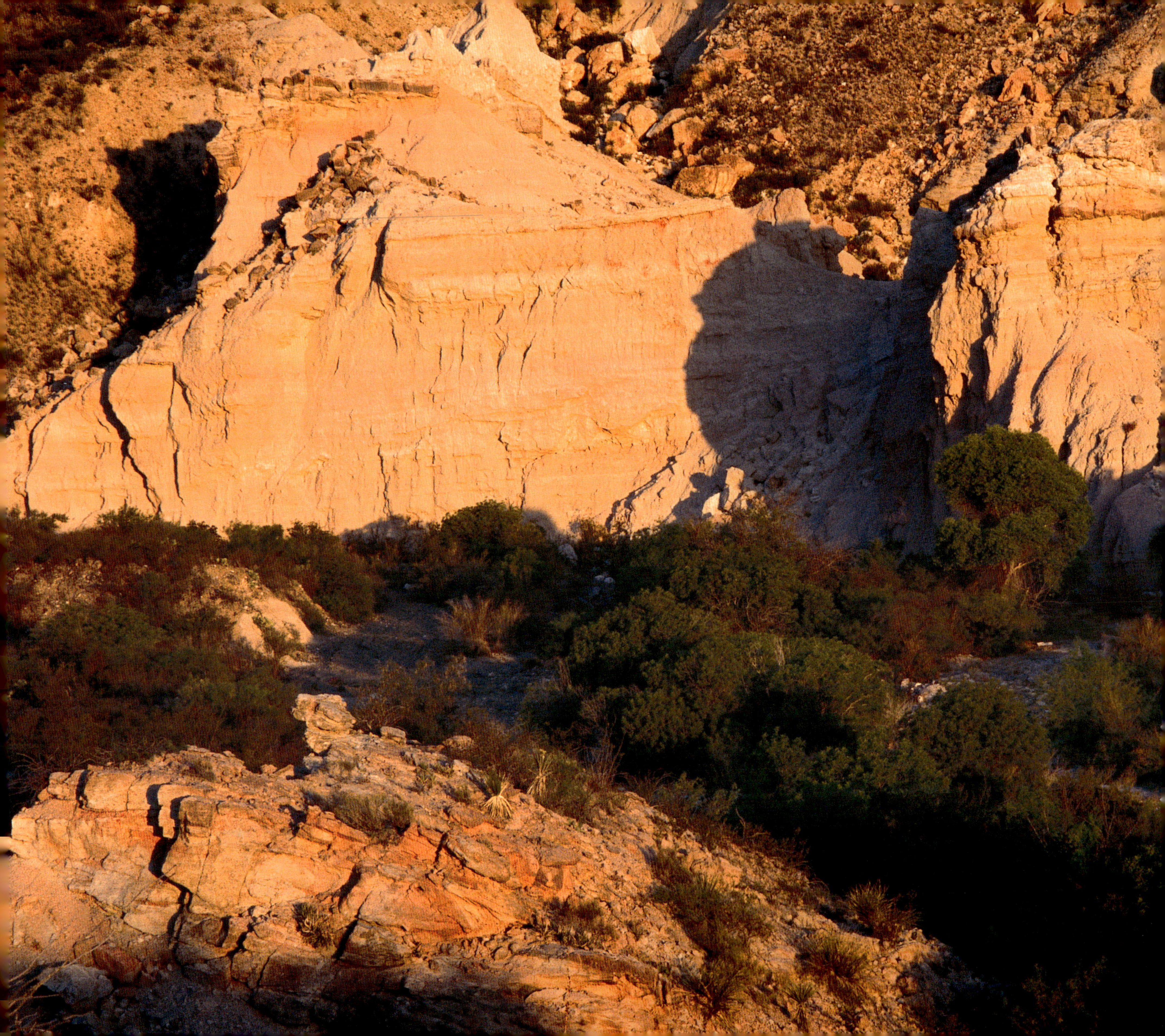

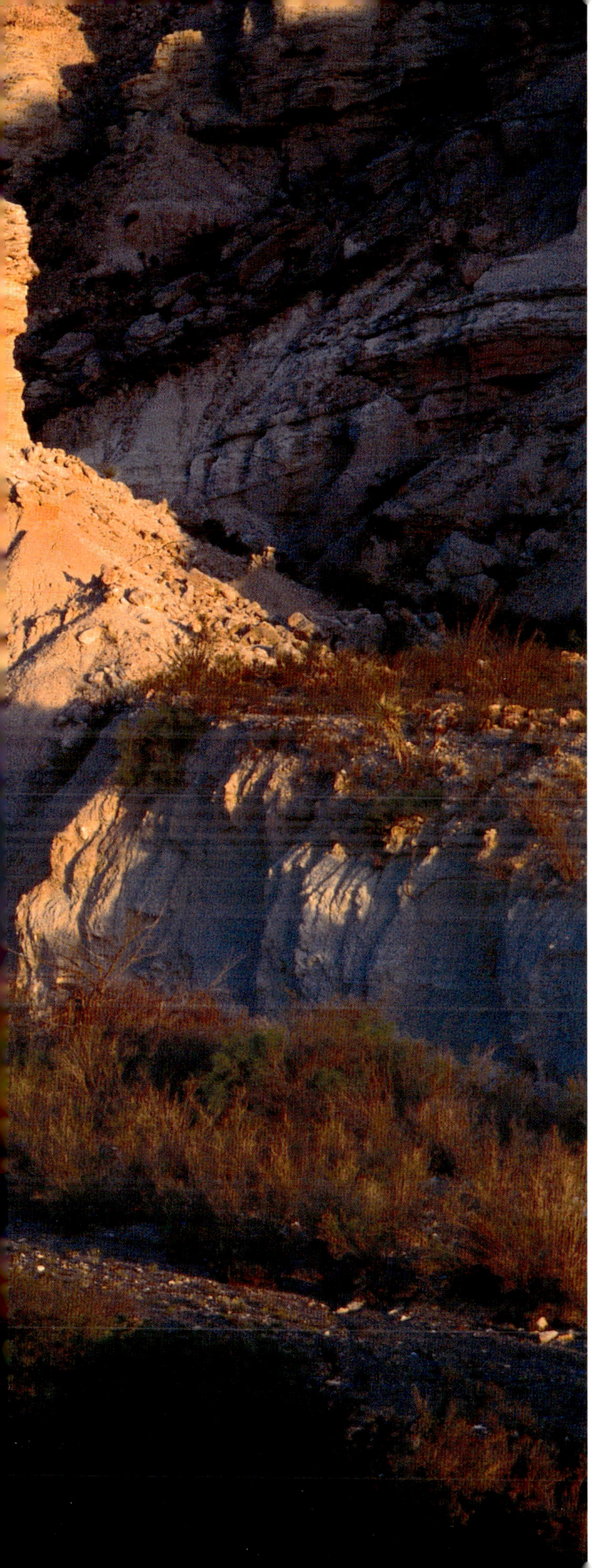

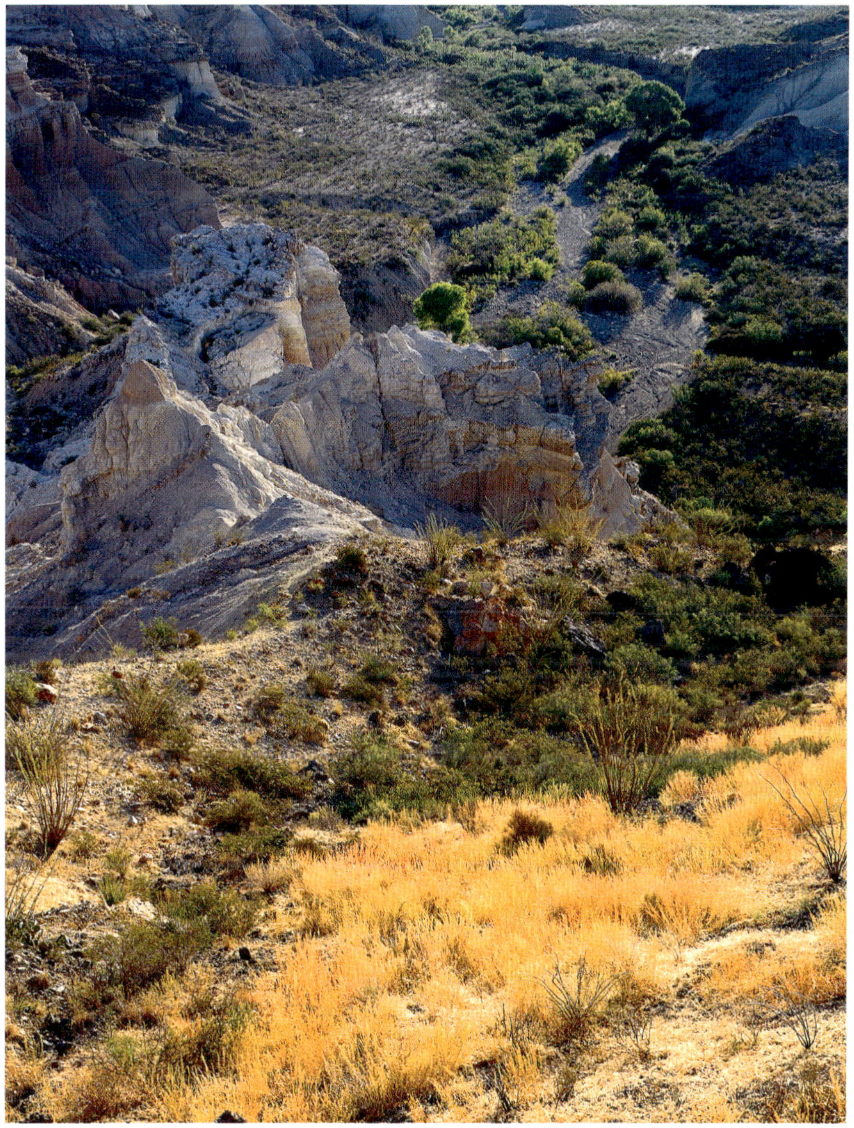

Opposite page: Volcanic ash beds in early morning sunlight.
This page: Early morning, Alamo de Cesario Canyon.

Apaches in the 1600s. The Apaches found the isolated mountains and canyons an excellent base from which to raid and plunder Spanish villages. Repeated Spanish military expeditions into the area had no effect—it was simply too vast.

The United States acquired Texas by annexation in 1845, but the Rio Grande boundary was not confirmed until the Treaty of Guadalupe Hidalgo was signed in 1848. To Americans, the Big Bend area remained an unknown and fearsome country. The first American to penetrate this forbidding land—passing not far from the Dalquest Site—was the redoubtable Captain Jack Hayes of the Texas Rangers, a short, pugnacious man full of reckless enterprise. In 1848, Hayes announced plans to find a route through the Big Bend, following the river, which would unite the markets of New Orleans and San Antonio with Chihuahua City. They left San Antonio with great fanfare and entered the Big Bend country from the east in high spirits. Almost immediately, Hays and his men became hung up in canyons, blocked by mountains and deep canyons at every turn.

The Apaches pilfered their horses. They were terrified, feeling they were being constantly watched. Their map was worthless. Food ran short. They found neither game nor wood. The men ate their mules and when those ran out subsisted on boiled beargrass. The expedition's doctor ingested poisonous plants and ran into the hills raving mad, never to be seen again. But Hays persisted and made it to San Carlos and on to Presidio del Norte, where he remarked laconically to Ben Leaton that the trip had been "fraught with more difficulties than anticipated."

Hays was only a few years ahead of his time. Others soon opened the very trade route he had sought, finding a way through the Big Bend along what had once been the old Indian Salt Trail skirting Cibolo and Alamito Creeks. This became the Chihuahua Trail. Enormous Chihuahua wagons loaded with luxury goods from New Orleans were soon crossing the Big Bend close to the Dalquest Site.

Meanwhile, American explorers had stumbled across the Great Comanche War Trail, which also bisected the Big Bend. For more than a century, during the full moon in September (the "Comanche Moon") the Comanches had left their homelands in the high plains of Texas and Oklahoma and cut south through the Big Bend to raid villages in Mexico. The Comanches returned from these expeditions in early January loaded down with captive women and children, large herds of stolen livestock, and other spoils of war. According to early American explorers, the trail was an astonishing sight: in places it was more than a mile wide, littered with the bones of animals and captives and "so well beaten that it appears that suitable engineers had constructed it." The Comanches marked good campsites along the trail by pinning a sapling to the ground, and at least one of these Comanche marker trees can still be seen in the Big Bend. It is now an ancient oak bent in a bow with all its branches growing upright. Today this remarkable trail has all but vanished to the naked eye, although it is being retraced and mapped by researchers from the University of Texas.

The Dalquest Research Site, by chance, occupies a strategic position almost directly between the Chihuahua Trail and the Great Comanche War Trail. While no formal archaeological surveys have been done of the Dalquest area, there can be little doubt it was an important hiding place for Apache raiding parties and perhaps other Indians. Indian artifacts have been found on Dalquest peak, at Fern Spring, and in some of the almost inaccessible and hidden canyons of the site. There is permanent water and plenty of wild game. Dalquest peak (once

Opposite page: Alamo Spring.

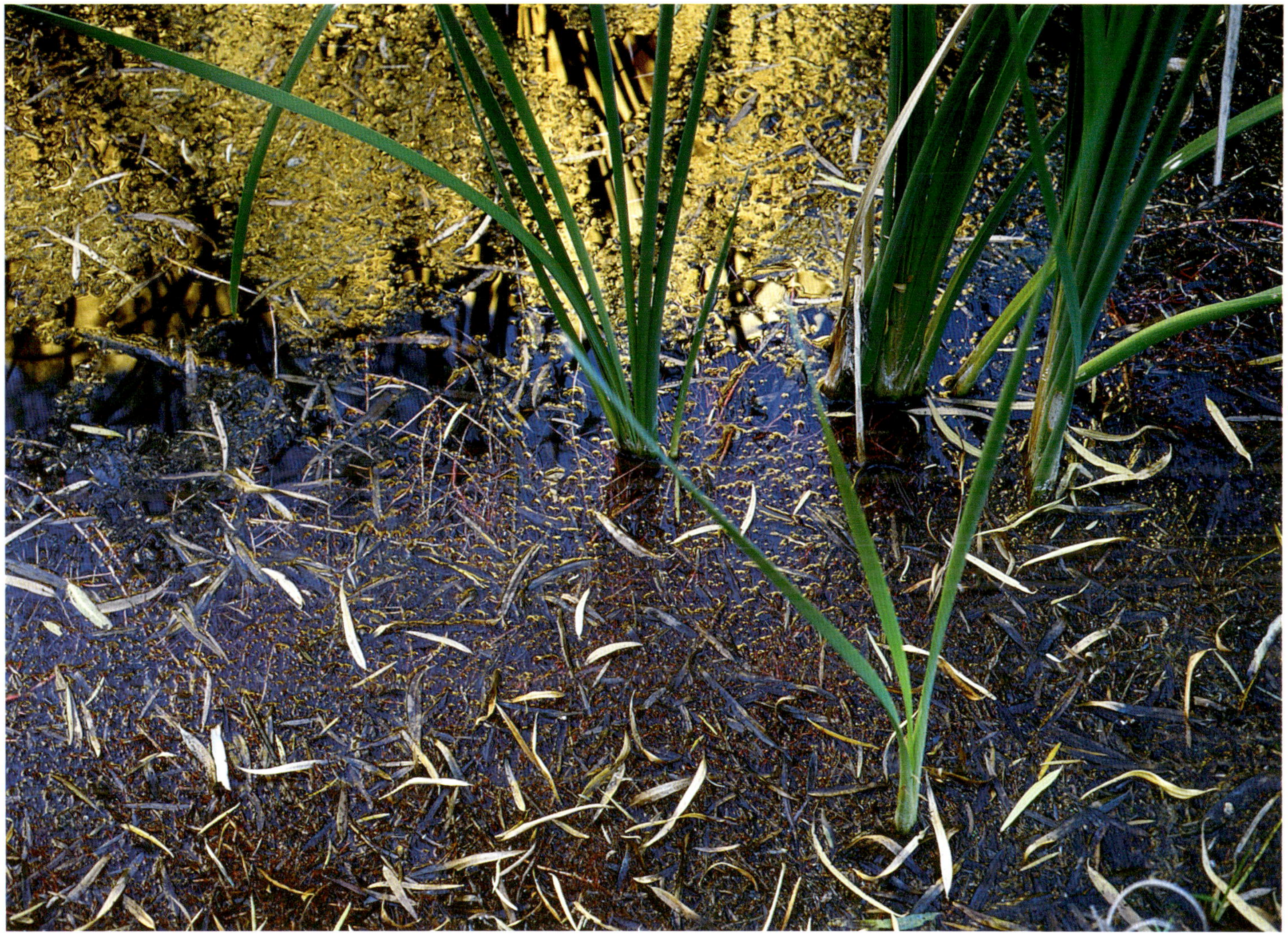

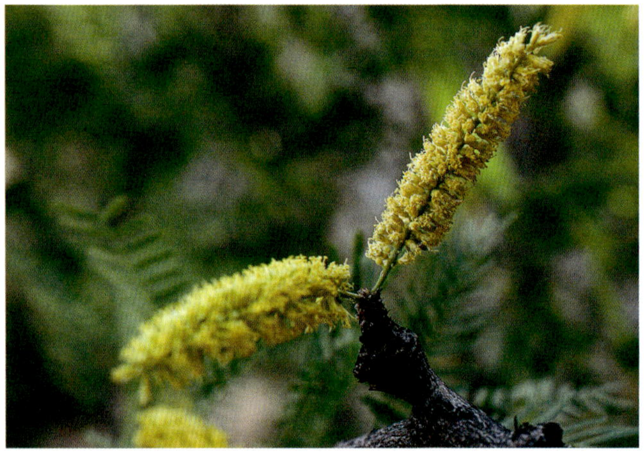

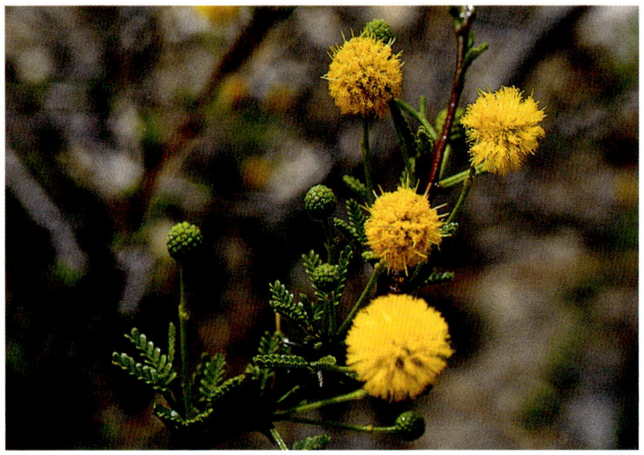

This page, above: Flowering mesquite.
Below: Acacia, *commonly called catclaw.*
Opposite page: Magnification of yellow rocknettle leaf.

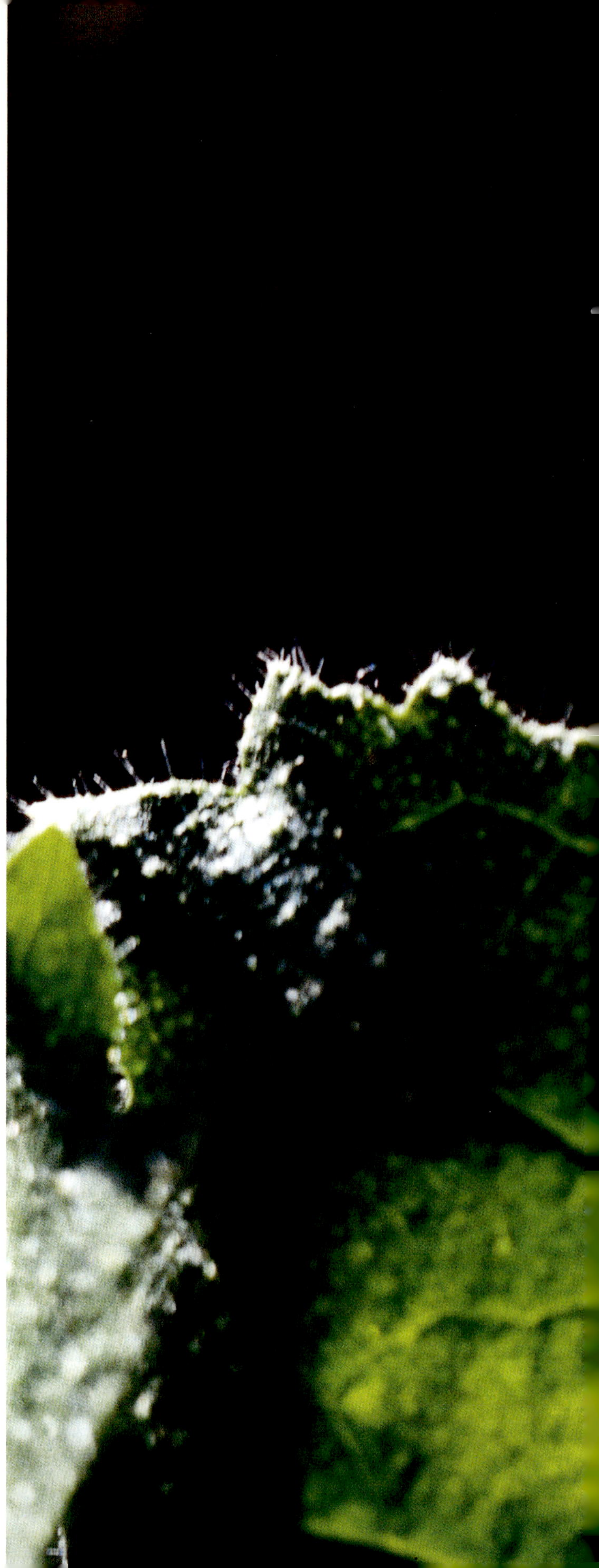

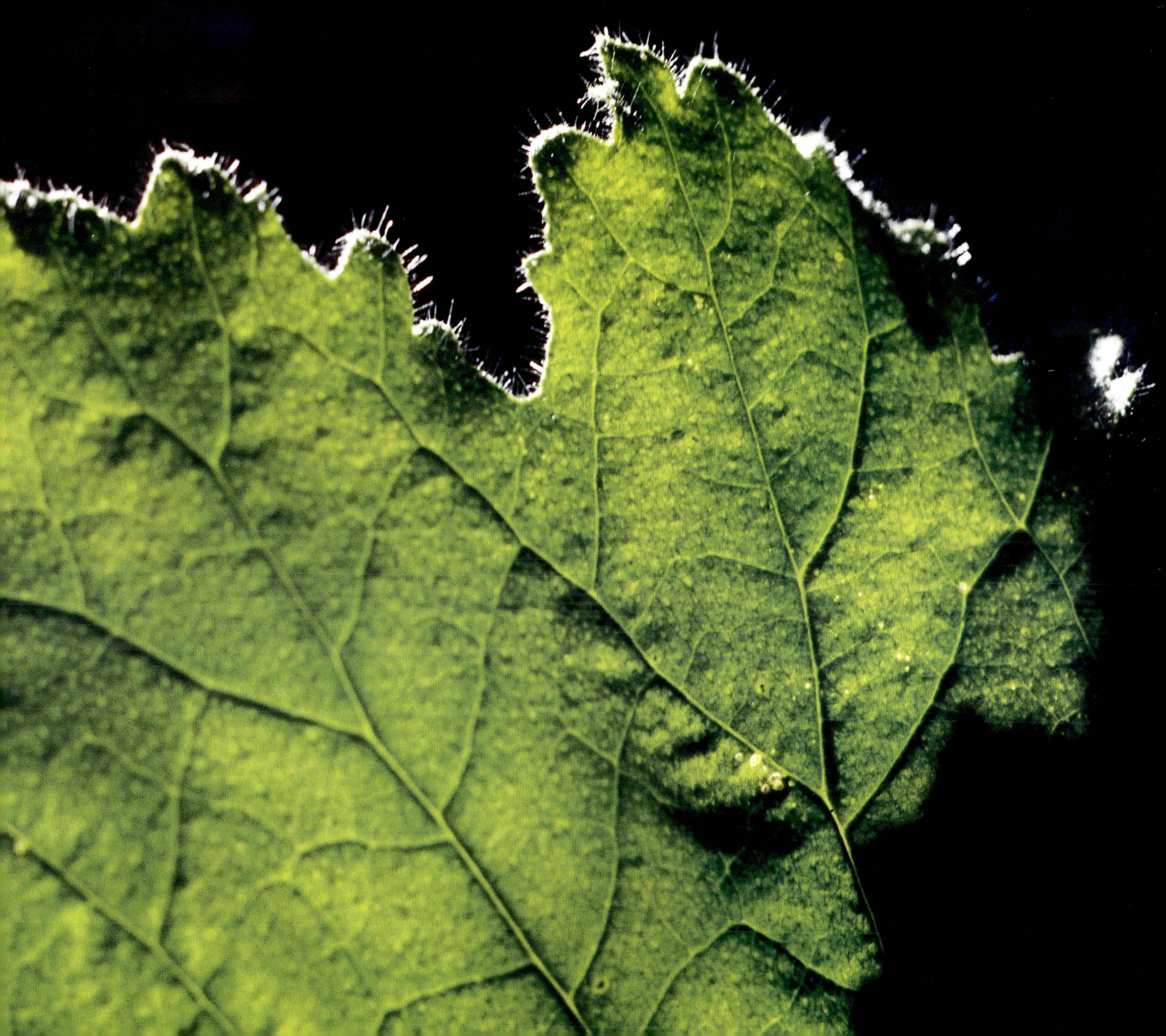

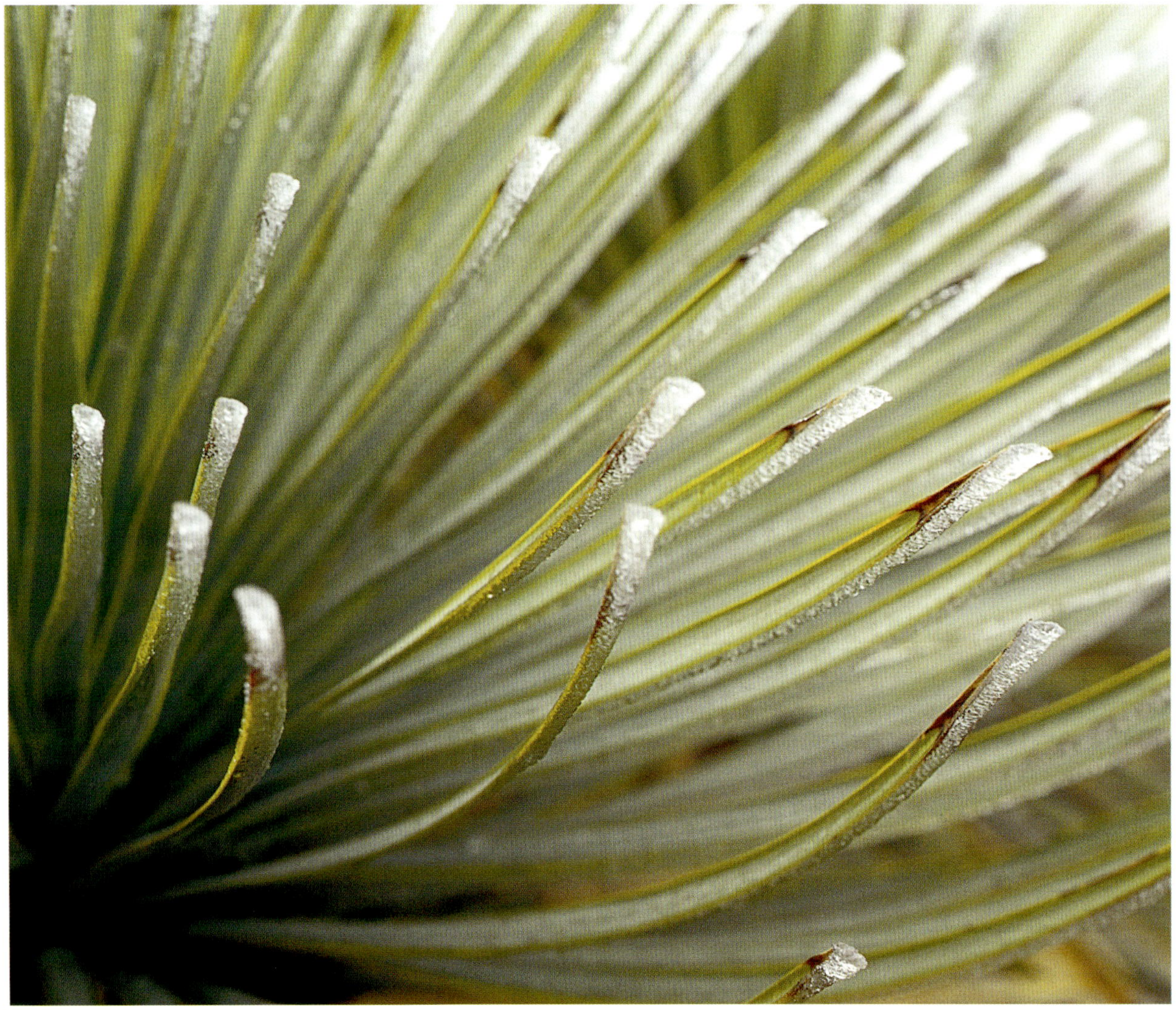

This page: Ice-covered yucca.
Opposite page: Desert grass along South Fork,
Alamo de Cesario Creek.

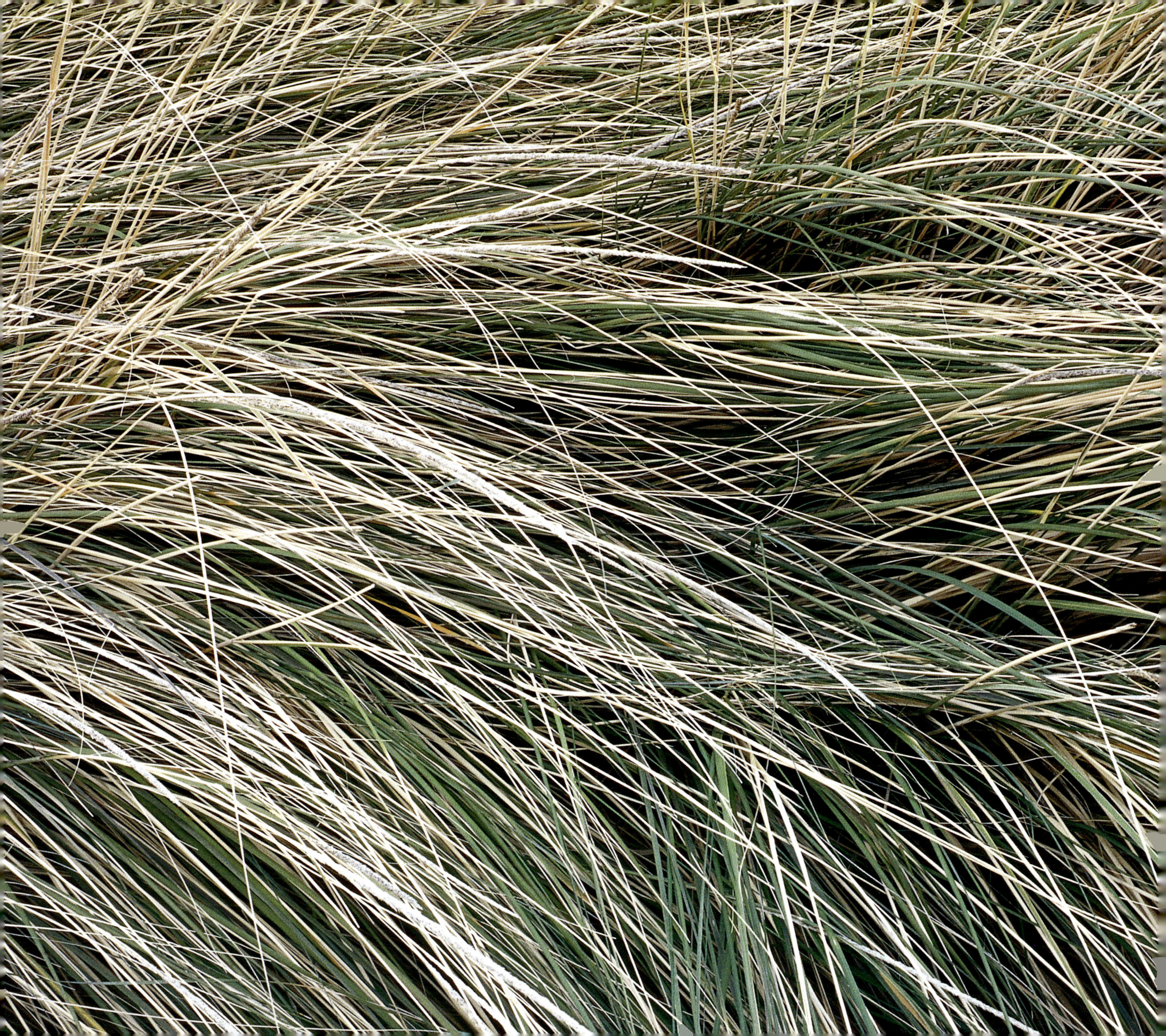

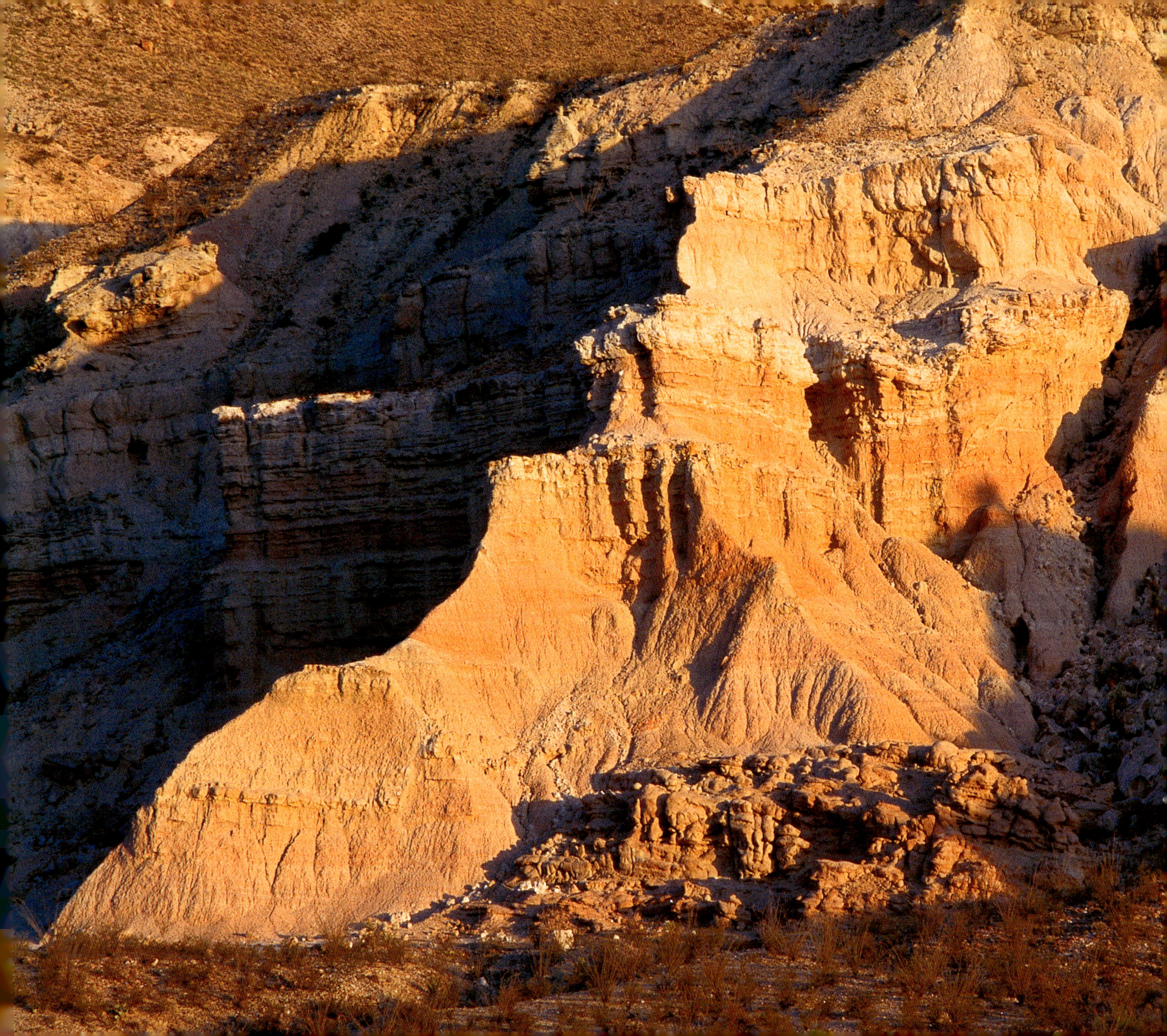

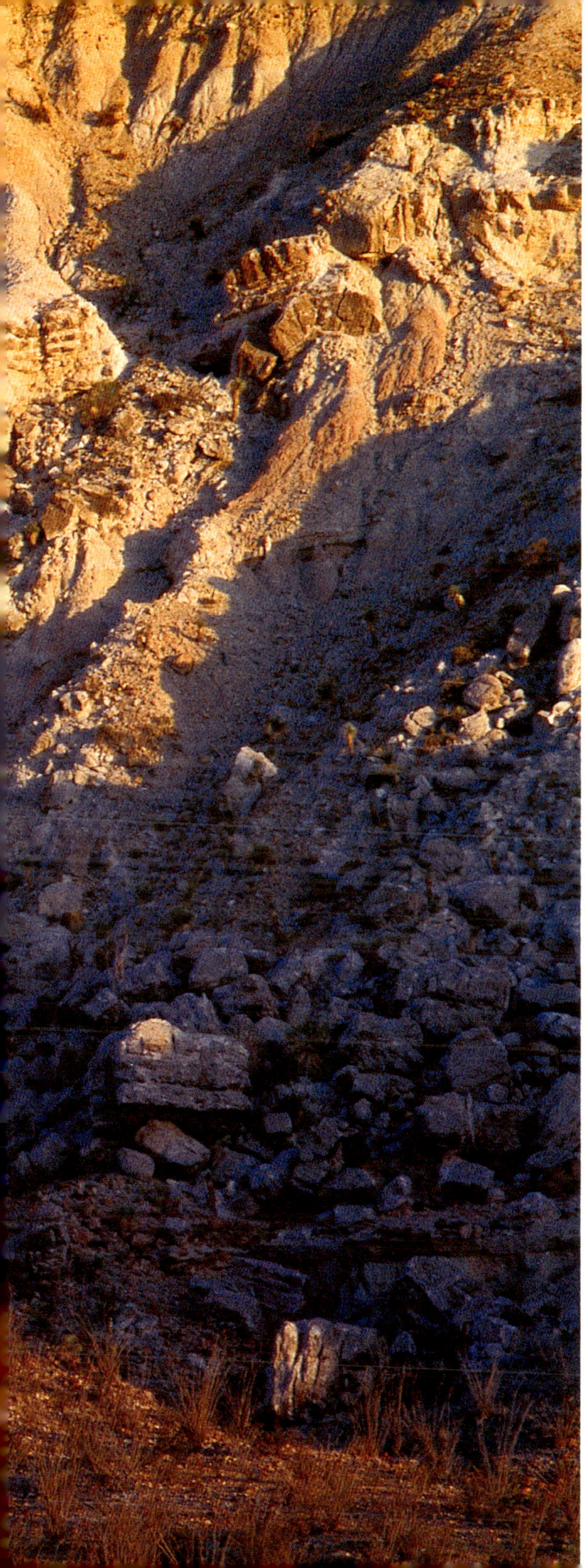

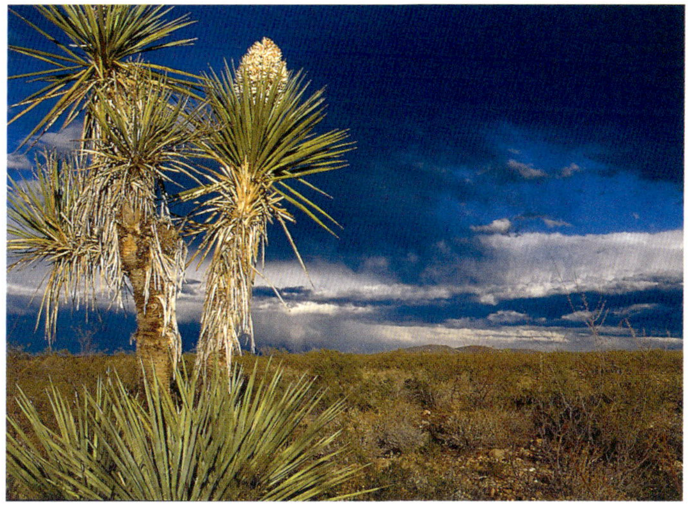

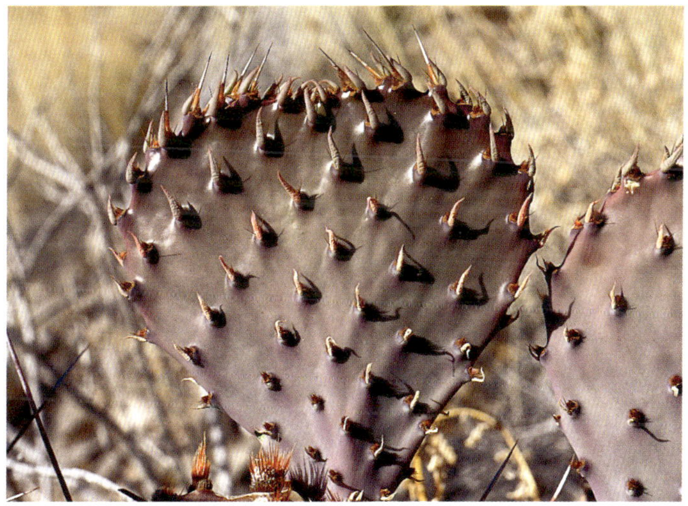

Opposite page: Volcanic ridge in Devil's Graveyard.
This page, above: Blooming yucca.
Below: Developing prickly pear pad.

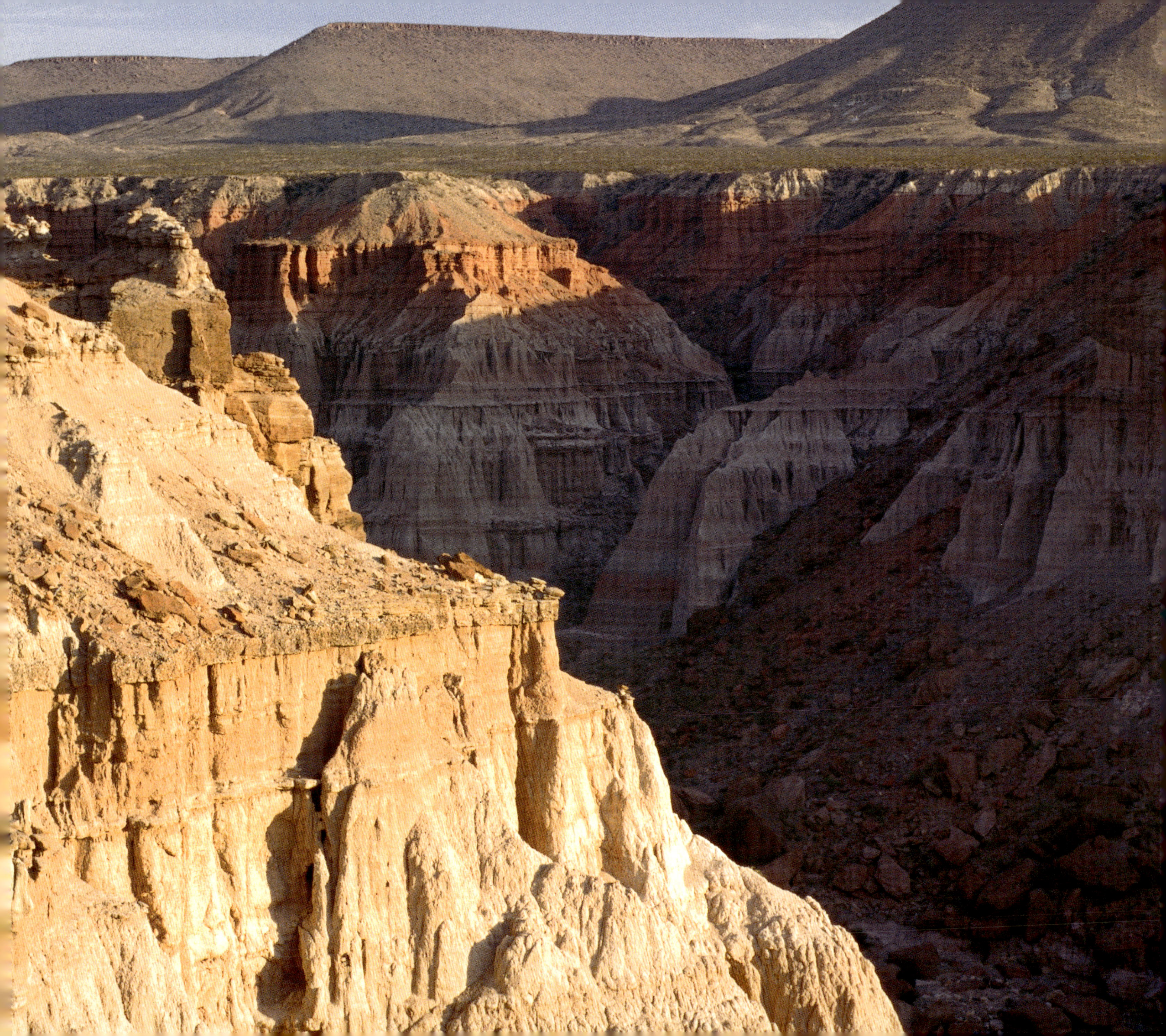

called Lookout Mountain) commands a sweeping view of the Great Comanche War Trail to the east, and it may have been possible to see parts of the old Chihuahua Trail from the same lookout. The area even includes a deposit of flint-like stone the Indians mined to flake into arrowheads and knives.

In the 1870s, the Big Bend area as a whole became a refuge for Apaches fleeing their reservations in New Mexico and Arizona, where corruption, drought, disease, idleness, and degrading conditions made life unbearable. Things reached a crisis in 1879, when the Apache chief Victorio fled the reservation with a force of warriors and swore to "make war forever" on the United States. Victorio worked his way southward into West Texas, where he inspired many Mescalero Apaches to rally to his cause. Using the Big Bend as a base, they stole livestock, killed cowboys and sheepherders, and drove away survey teams that were mapping a route for the coming railroad. Victorio did not limit his attacks to Americans. He had a special hatred for Mexicans, and he swooped down on villages south of the border, wreaking havoc. In one of the most famous massacres in Mexican history, Victorio lured a group of Mexican volunteers from the town of Carrizal, west of the Big Bend, into the Candelaria Mountains, ambushed them, and then ambushed those who came to their aid; by the time the dust settled, every adult male in Carrizal was dead.

When Victorio tried to return to the United States after the Carrizal massacre, the commander of Fort Davis, Col. Grierson, and a force of Buffalo soldiers cut him off in the Big Bend and drove him back into Mexico. In August, Grierson heard from scouts that Victorio was once again on his way to the U.S. border. Grierson realized that the Indian warrior would have to stop at Rattlesnake Spring to water his horses. With no time to spare, he pushed his Buffalo soldiers on a brutal twenty-four-hour march of sixty-five miles and reached the spring just ahead of Victorio. He hid two companies of 10th Cavalry on each side of the ravine facing the spring. When Victorio and his warriors arrived, they were caught in crossfire and cut to pieces. The survivors fled back into Mexico, shattered and demoralized. The pursuit was taken up by Mexican troops, who trapped the Apaches on an isolated knoll in the desert and proceeded to pour fire into it. Around midnight, the Mexicans heard the death song of the Indians rising from the knoll; Victorio had been killed and Apache resistance in the Big Bend had been broken.

Now the building of the railroad went full speed ahead. In 1881, seventeen cars of iron and forty cars of ties were being laid each day with the labor of Chinese workers. Due to a lack of surface water in the Big Bend, wells had to be dug every twenty or thirty miles to supply the boilers of the steam engines. Towns sprang up around these wells and became known as tank towns. Marfa was one such tank town. It was named, so the story goes, by the wife of the chief engineer, who was reading Dostoevsky's recently published novel, *The Brothers Karamazov*; she plucked the name of an elderly servant, Marfa, out of the novel.

Even though Victorio was dead, there were still some renegade Apaches roaming the Big Bend, desperately clinging to their old way of life. One settler recounted an interesting tale of the Indians' first encounter with the railroad. In 1881, a ragtag group of Apaches killed and butchered for food a strawberry roan milk cow named Old Babe that belonged to a local rancher. A posse of Texas Rangers set off in pursuit. Near Kent, the posse spied the Apaches in time to witness a sad encounter. The Indians had come across the newly laid railroad tracks,

Opposite page: Red canyon, looking west toward Bandera Mesa.

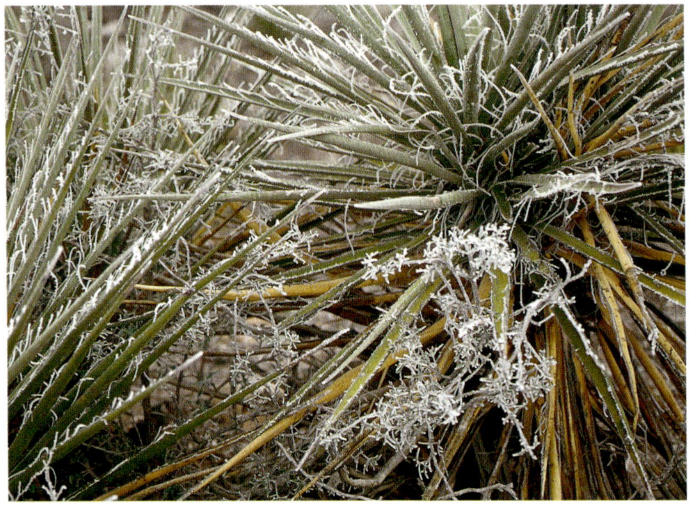

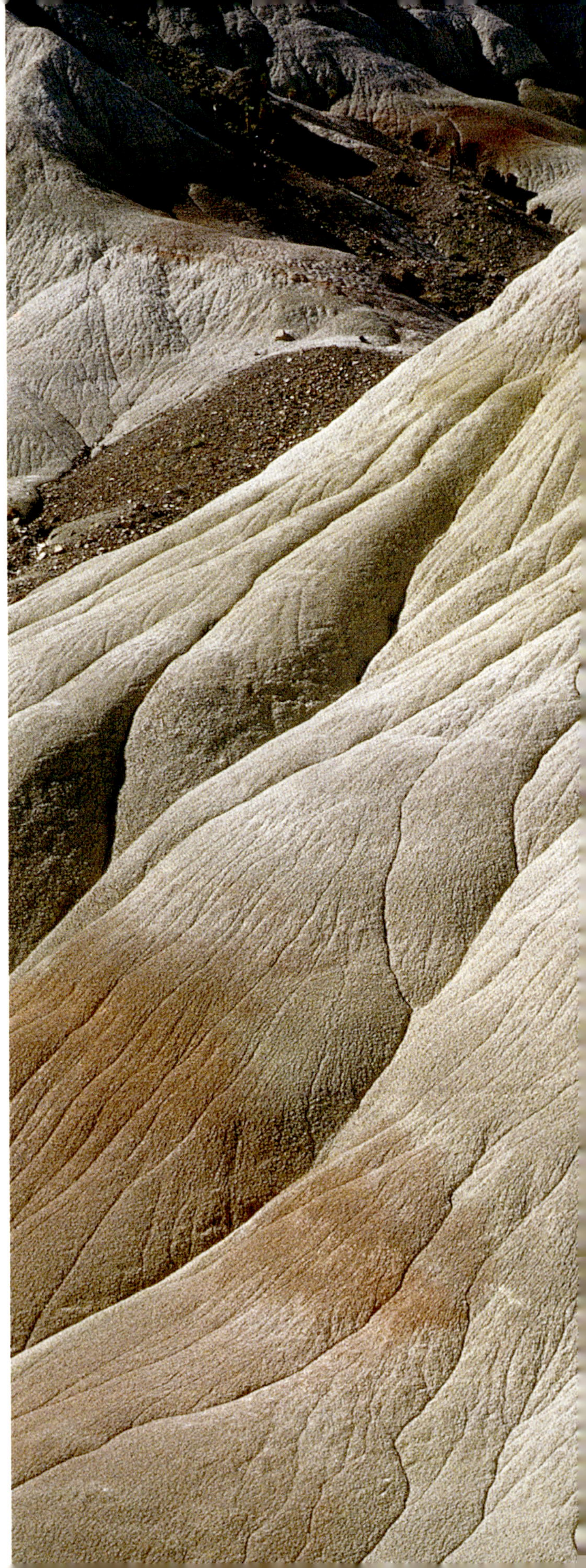

This page: Ice crystals on yucca.
Opposite page: Water-sculpted volcanic ash.

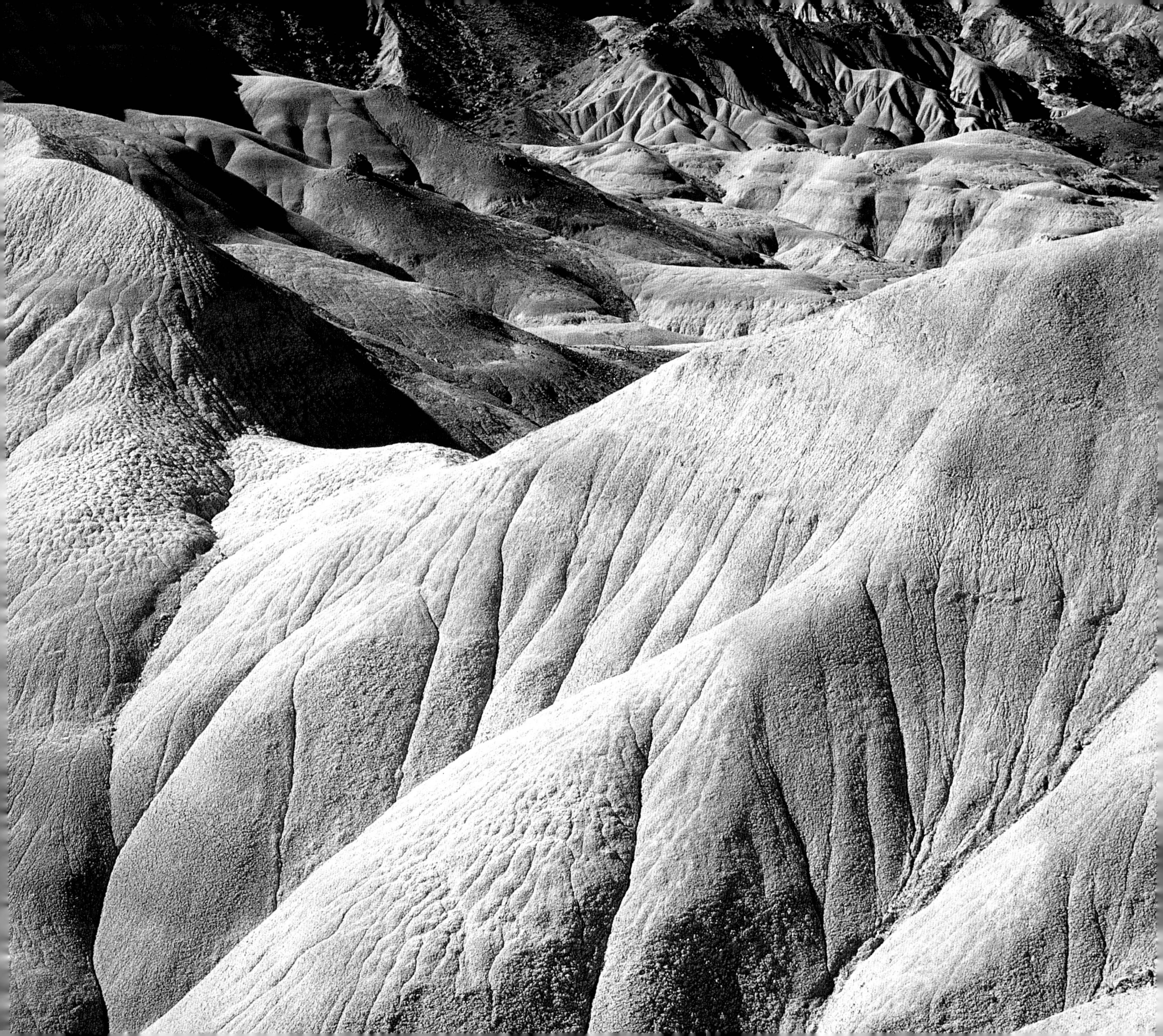

and were utterly baffled by the fresh earthen embankment extending to the right and left as far as they could see. The Indians broke into war whoops and galloped back and forth, brandishing their weapons, as if they could somehow drive this strange excrescence from the landscape. In January of 1882, the sledgehammer rang down on the last spike joining the Southern Pacific with the Galveston, Harrisburg, and San Antonio Railroad, opening the Big Bend. A newspaperman at Fort Davis declared, "We are no longer the frontier."

The completion of the railroad had a far-reaching effect on the land that would become the Dalquest Site. In order to encourage the building of the railroad, the state of Texas passed a law giving the railroad companies sixteen sections of land for every mile of track they laid, with the proviso that the railroad survey another sixteen sections for the state. (A section is a square mile of land.) The state portion of this land was to be used to finance the public school system, and it became known as school land. Millions of acres of school land and railroad "scrip" land were surveyed, usually checkerboarded in alternating sections. The two sections that make up the Dalquest Site were part of this checkerboard: one piece was deeded to the railroad, the other designated school land.

The railroad immediately resold its land holdings. On January 8, 1884, the Texas and Chihuahua Railroad Company sold one section of the future Dalquest Site to a man named Havard. The other section remained school land.

Papa Tom Rawls

WITH THE JOINING OF THE RAILROADS AND THE FINAL EVICTION OF THE Apaches, settlement proceeded, and by the turn of the century most of the great ranches that would dominate the area had already begun to form. The land encompassing the Dalquest Site was owned by a legendary cowman by the name of Thomas H. Rawls, "Papa Tom" to young people. Born in Georgia in 1868, Rawls came to the Big Bend seeking land and fortune. He started out working for a man named McGuirk for $20 a month, but he was obsessed with owning a ranch of his own, and after six years he had saved up enough money to buy his first spread, the Alamo Cesario Ranch.

Rawls was a small, leathery man who chewed tobacco and drank sotol, the potent extract of the sotol plant, even in the days when it was illegal. He kept a keg of it conveniently buried near his house. People said he "drove a car like a team of mules," urging it along with curses and imprecations, hollering for everyone to clear out of his way. He kept a coffee can in the middle of his living room floor to serve as a spittoon and "didn't care if he missed it." He loved children and romped with them like a little boy himself. But more than anything else he loved wheeling and dealing in ranch land. He sold the Alamo Cesario and bought the San Jacinto Ranch. He acquired the two sections of land making up the Dalquest Site when he bought the Bandera Ranch. (It reportedly cost him a couple of dollars an acre.) He bought the Holquin, Alazan and Tascatal ranches—all the while swapping, buying and selling bits and pieces as he went along, forever in search of better pasture.

Rawls and the other big stockmen were greatly aided by the rush of settlers that began pouring into the country. The State of Texas was selling land at low prices to encourage settlement. In order to prevent the formation of large ranches, the State did not allow any one person to buy more than four sections of state land. The settlers snapped up the cheap state land in small parcels. They quickly became known as "starve-outs." Virtually all of them came from wetter climates farther east, where a square mile was a kingdom and four square miles a veritable empire. In the Big Bend, however, a square mile of land would barely support a few horses, a brace of mules and a milk cow. If the section didn't have year-round water (and most didn't) it was a death trap. When the rains were good, so many homesteads sprang up that one cowman complained he was never out of sight of them. The prosperity was a cruel illusion: when drought came, as it inevitably did in that country, the starve-outs lost their livestock and were forced to sell their hardscrabble outfits. For years afterwards their pathetic rock and brush houses could be seen crumbling back into land from whence they had risen. In the process the starve-outs converted much public land to deeded acreage that was repurchased by the big cattle outfits, assisting in the creation of the big ranches the State sought to prevent.

Through buying up starve-out land, Papa Tom Rawls greatly expanded his holdings. At one point he ran livestock on 60,000 acres with 33 windmills and hundreds of miles of fence. His quality horses were in demand from as far away as Louisiana. He built a rock house on the Alazan Ranch and settled down with his wife, Dolly, a famously independent woman who once split the skull of a bobcat with a stick.

The Mexicans called Rawls Don Tomás. Once a Mexican foreman stole fourteen of his horses and fled to Mexico, bringing them into Pancho Villa's camp. Papa Tom strapped on his big guns, paid a cordial visit to Villa, and asked either for payment or the return of the horses. Villa liked the wiry, determined gringo and paid Rawls a good price for them, adding a small gift, traditional in the Mexican way of doing business, of three pairs of silver dice and a ring in the shape of a rattlesnake. Rawls continued to sell horses to Villa during the years of the Mexican revolution.

Just before the First World War, a young man named Pearl Andrew Jackson and his brother Mark bought the Bandera Ranch from Rawls, thus acquiring one section of the future Dalquest Site. Mark stayed at the ranch and took care of the cattle and horses the brothers were slowly acquiring, while Pearl, with mules and a fresno (a heavy dirt scraper), hired himself out to build dirt tanks, earning cash to support the ranch. The other Dalquest section was still school land, and therefore technically public, although by local custom the ranchers ran their cattle gratis on school land and took umbrage at any attempt to collect fees. Eventually Texas cracked down on the free use of school land but gave ranchers the opportunity to buy it. In 1934, Pearl and Mark acquired the second section of the Dalquest Site. Pearl bought out Mark's share and came to own the ranch outright.

In 1941, Pearl Jackson, by then middle-aged, acquired a family when he married Sophia Glascock, who had a daughter, Lee, from a previous marriage. The Bandera ranch would eventually go to Lee, the last owner of the land before Walter Dalquest.

Lee Glascock Bennett still lives in Marfa. She taught history for many years in the high school and at Sul Ross State University. Lee arrived in Marfa in 1925, under unusual circumstances. When she was three years old, living with her family in San Antonio, she fell dangerously ill with whooping cough. The illness resisted all efforts at treatment. The doctor finally became so alarmed that he told her parents that to save her life they had to get her out of San Antonio to a drier climate farther west. Lee's parents loaded up their four-door Buick and headed toward El Paso, where they had relatives. They traveled thirty to fifty miles a day on dirt roadways, spending nights in various towns along the way. Every night, Lee coughed until dawn.

One night, while they were staying at the old Alta Vista Hotel in Marfa, Lee slept through the night without coughing. The next morning her father announced that this was where they were going to move. His wife protested that they should continue on to El Paso, but he was adamant: this was where Lee had stopped coughing, and this was where they would stay.

Lee's father died in 1936, after a long struggle with cancer. A few years later Pearl Andrew Jackson came courting Lee's mother, Sophia. Jackson had come a long way from his days of doing dirt work with a fresno. In his mid-fifties, he had established himself as a successful rancher with good cattle, fine horses and a reputation as a hardworking, honest, gentle man. Evidently, Jackson had never thought that he would have an heir, but at one point in his wooing of Lee's mother, he stated, "I want Lee to have my ranch." Eventually Lee did inherit the ranch, which consisted of 21,000 deeded acres, including the two sections of the Dalquest Site, which at that time was called the Falls.

In 1948, Lee Glascock married a man named Murphy Bennett. They moved out of Marfa in 1951 with their two young children to live at the isolated Bandera ranch headquarters, more than sixty miles south of town, thirty miles of which were on a bad road frequently swept by flash floods. The ranch house (which is presently uninhabited) lay in a pleasant draw a few miles north of the Falls. At the time many of the ranchers were switching from cattle to sheep, which seemed to make economic sense. Lee and Murphy Bennett did the same.

And then the infamous drought of the 1950s hit. Hungry sheep, unlike cows, pull up grass by the roots—a fatal difference. The sheep peeled the land. When the grass was gone Lee and Murphy kept their sheep alive by "knocking" sotol and chopping the heart and leaves with a hammer-mill mounted on the back of a wagon. During those tough times Murphy never lost his sense of humor. Lee remembers him coming in one day and announcing: "Well, they've eaten all the grass, the brush and the sotol, and now they're starting in on the small rocks."

Even so, Lee Bennett recalls those days as among the happiest of her life. "We were in the midst of the most terrible drought," she said, "twelve months without even a sprinkle. We were going broke and happy at the same time. We were well, we had each other and two wonderful little boys and we had hopes of better times." Their children were not yet old enough to go to school, but as they reached kindergarten and first grade age, Lee taught them at home from The Calvert Course. Murphy fixed up a sand pile for them to play in, shaded with a brush arbor where they had a swing and an assortment of small, dirt-moving vehicles. Murphy raised a string of beautiful quarter horses, and as soon as the children could walk they were put on horseback. The family went to town every two weeks—a major event. "We were broke, but our children didn't want for one thing, and we had experiences that could not be bought," Lee said. She remembers one morning when Murphy came rushing into the house, urging her to get the little boys and hurry to the pickup. They drove about two miles

Opposite page: Ocotillo surrounded by lechuguilla and prickly pear.

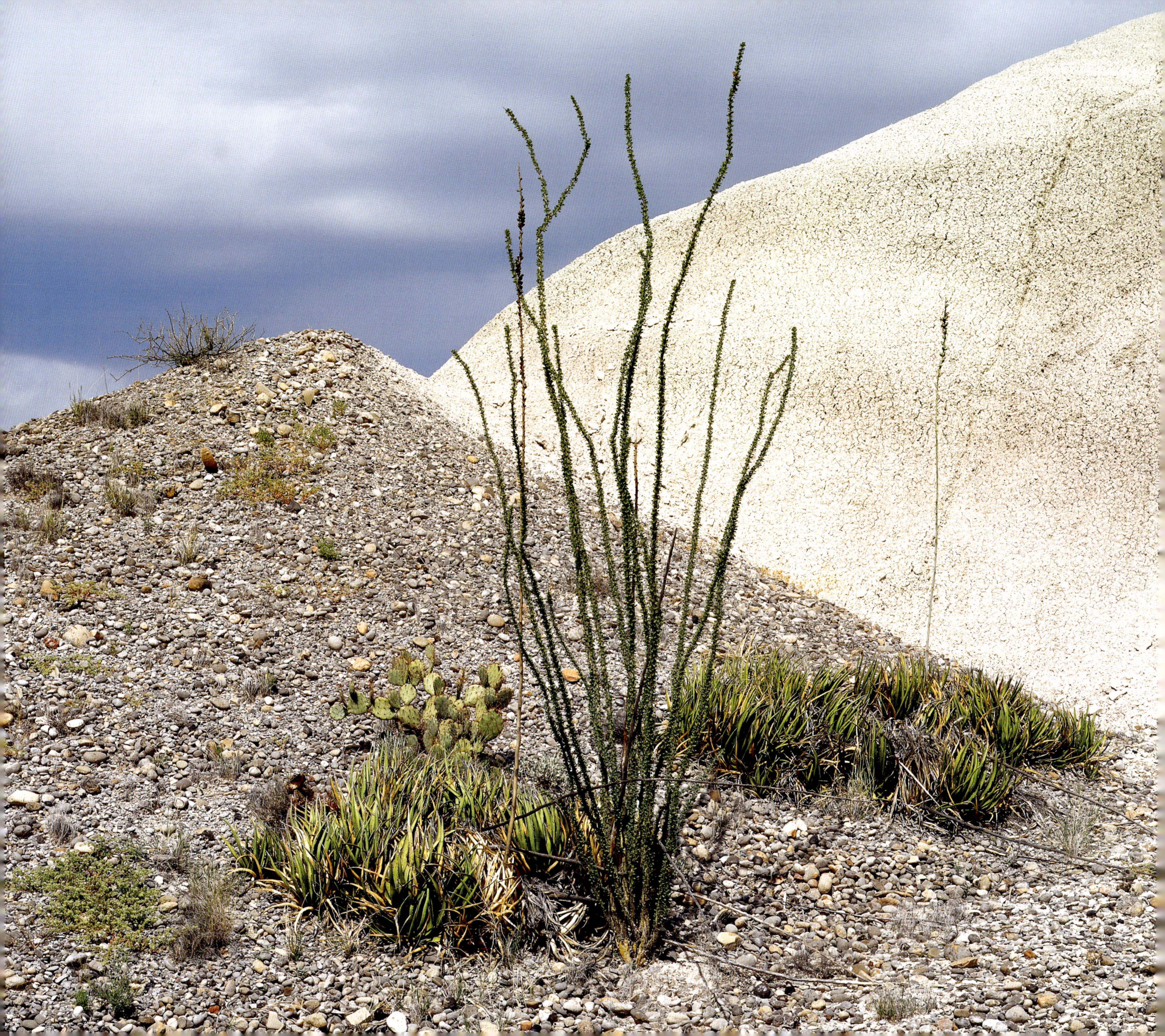

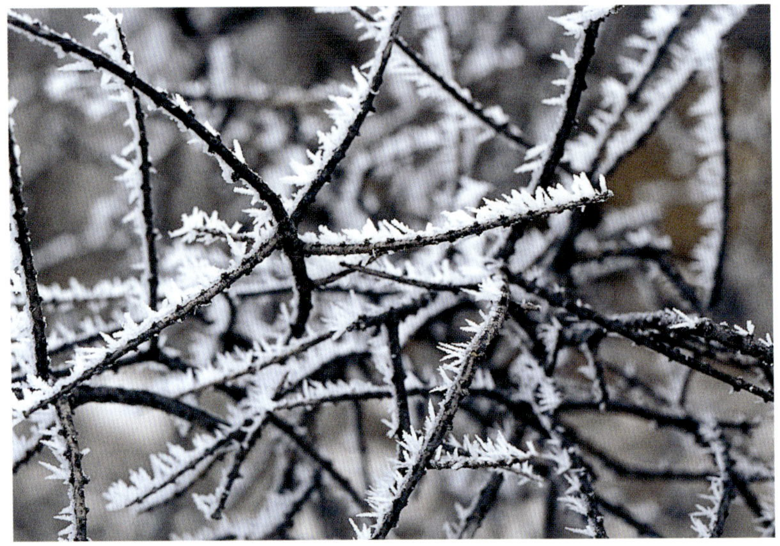

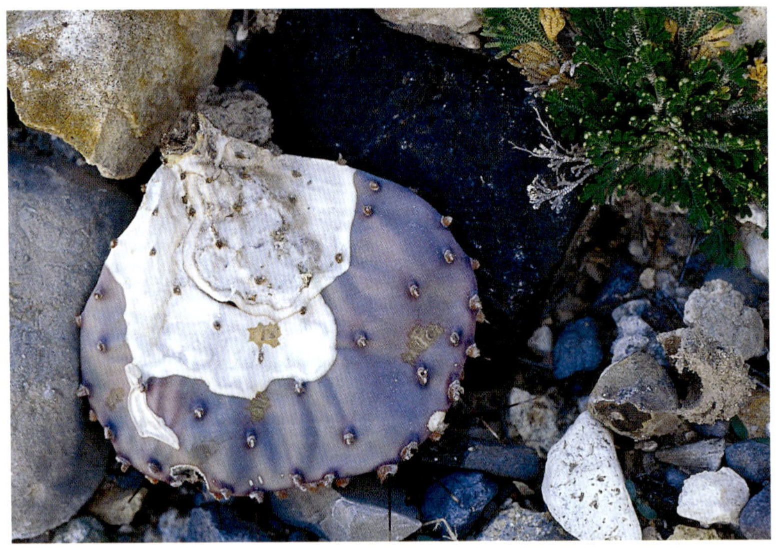

This page, above: Frosted catclaw.
Below: Dying prickly pear pad and resurrection plant.
Opposite page: Basalt lava flow with rim-lit tuff.

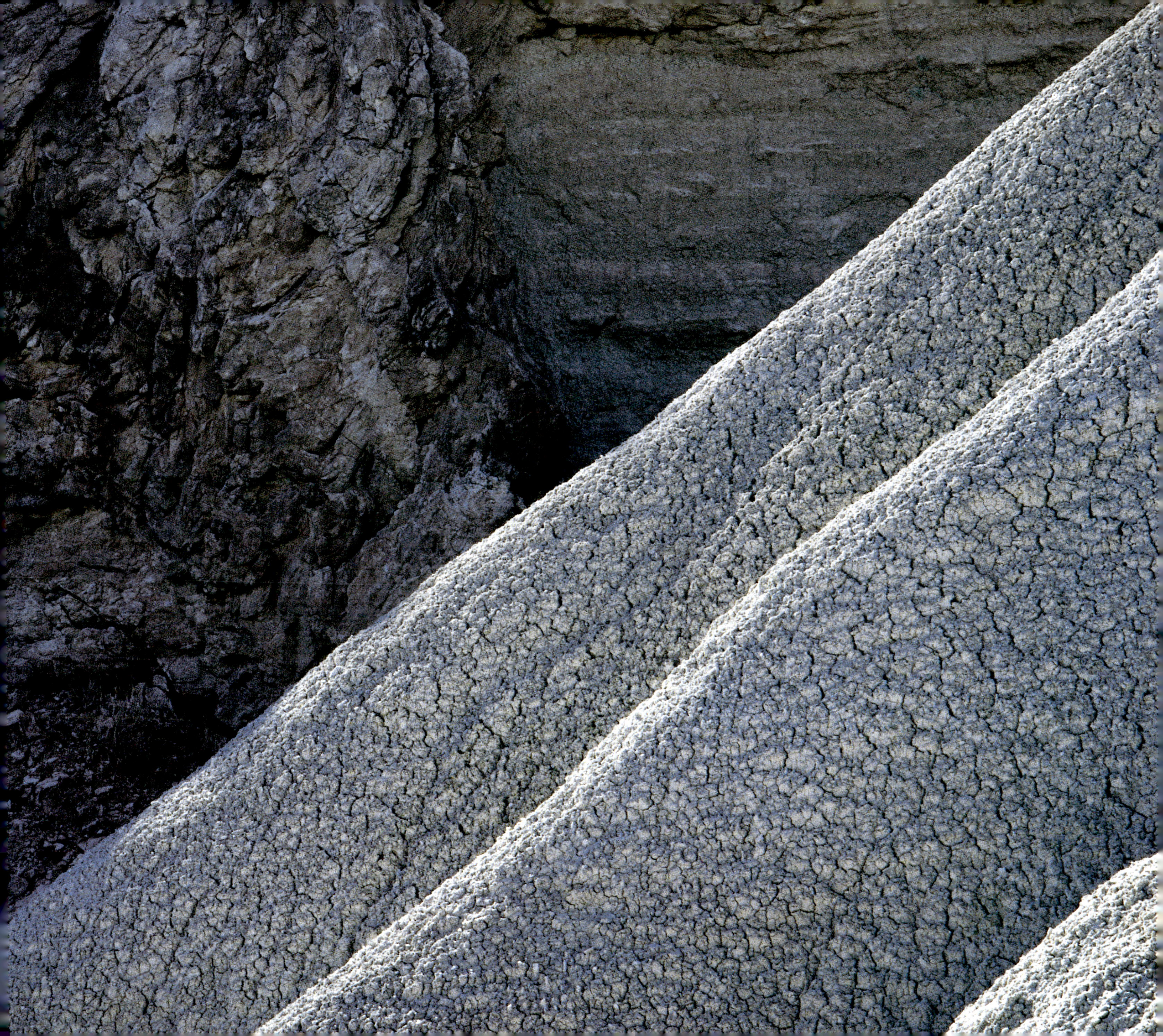

to a windmill and water trough near a hill. "The hill was about one hundred yards long and about thirty feet high, and it was totally covered with an undulating bright blue and gray blanket of thousands of migrating mountain bluebirds that were resting and watering at the site. Did we feel deprived? I don't think so."

While Lee and Murphy often went to the rim of the Falls, they rarely went down. Lee herself went down only three times—mostly trailing lost cattle. One time, Murphy took her down into the Falls on horseback, to have a picnic lunch on the banks of Alamo Spring. "It was just beautiful," Lee said.

In 1955, when their oldest boy reached second grade age, they moved back to town so he could attend school. They pulled the sheep off and let the land rest for seven years. During that time Murphy made a living working three or four different jobs. He bought and sold sheep and cattle, a career that he followed for an additional thirty years. He bought two used bulldozers, put operators on each and developed tank building and soil conservation diversions. In town, he maintained a clothes cleaning shop. Lee said, "He could work or mechanic a bulldozer and he could press fine silks. He was good looking, hard working, a good dancer, and oh! did he love horses!"

The Falls bordered the southeastern edge of the ranch, but it always seemed to belong to another world. There wasn't enough forage to put cows down there for year-round usage so it was left alone. "We tried our best to think of some way to use it," Lee said, "but we never could. It was just there. Nobody in Marfa even knew it existed. That land wasn't forgotten because nobody knew about it in the first place." It was not a place to tarry in. "During the summer if you saw a little cloud, it could develop quickly into a heavy rain. In no time you could hear the roar of water. Down there, when it rained, the water would roll eight-foot boulders along like marbles."

In 1968, Murphy talked to Lee about selling the Falls. She protested: "But it's so pretty down there."

Murphy replied: "Cows can't eat pretty."

So the site went up for sale. Walter W. Dalquest bought it, site unseen, mostly for herpetology research. It was an unusually lucky event. Dalquest happened to be one of the most celebrated biologists in the state of Texas, the first Hardin Professor at Midwestern State University in Wichita Falls, and an internationally known mammalogist and paleontologist. He was in a unique position to appreciate the true value of the canyon. Not only was it a geological laboratory, but it contained a unique and unspoiled ecosystem. Dalquest was a professor who believed his place was in the field with his students, and he often brought students out to the site to study its flora, fauna, geology, and paleontology. In 1996, four years before his death, he and his wife Rose bequeathed the site to Midwestern State University, to be preserved in perpetuity for scientific research.

The Dalquest Site has already become what Dalquest hoped it would be: a living laboratory for scientific research. It has already produced some spectacular scientific discoveries. A few years ago, Dr. Norman Horner, an entomologist at Midwestern, had a graduate student who discovered a new species of spider, which is under intensive study by scientists at the American Museum of Natural History in New York because it appears to belong to a family unknown to the Americas. The fossilized jawbone of a 50,000-year-old monkey was found at the site. Graduate students are engaged in research ranging from a study of bacteria in the water, a limnological study of the two permanent springs, and ongoing studies of the area's geology.

Opposite page: Dalquest peak at sunrise.

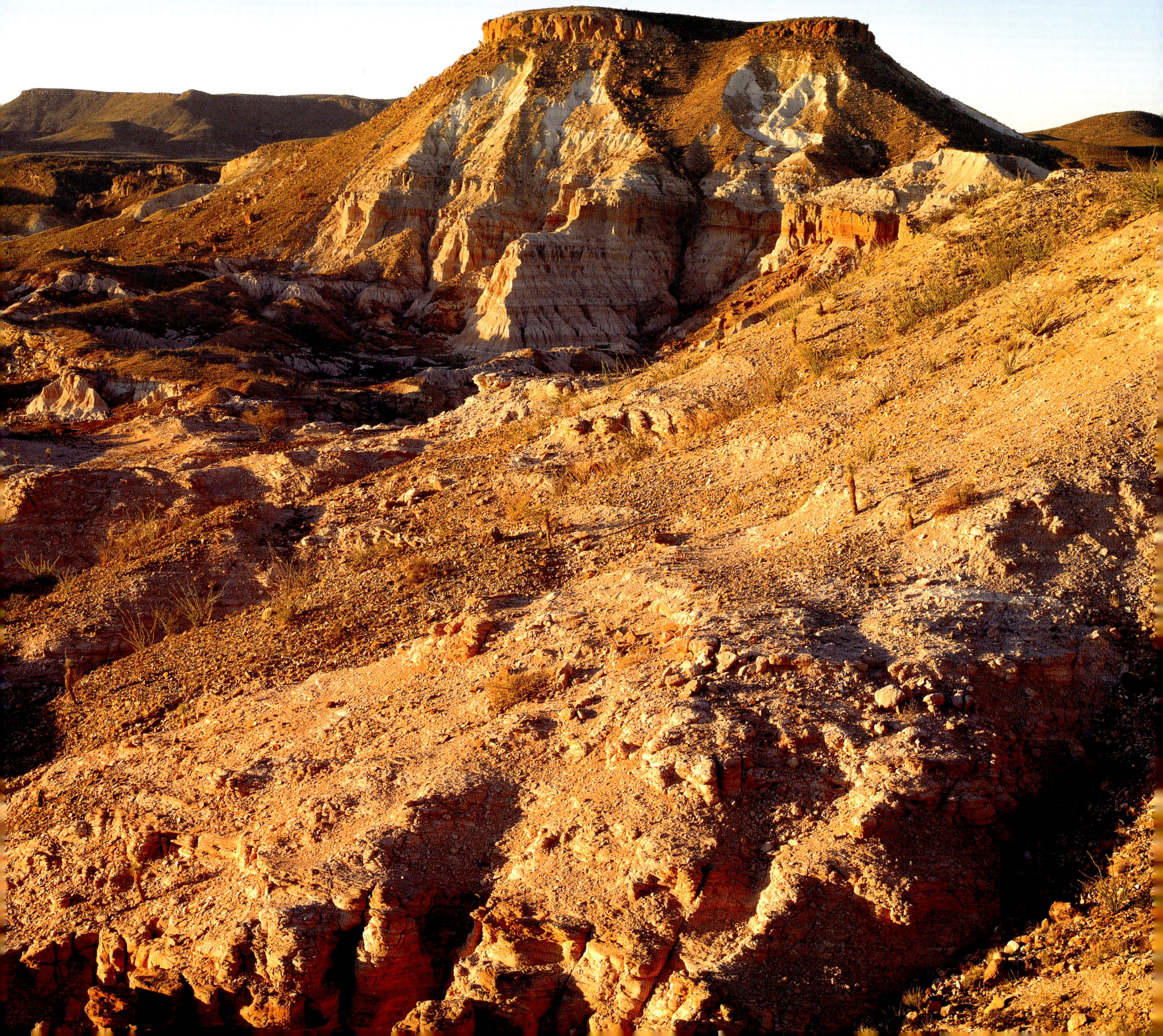

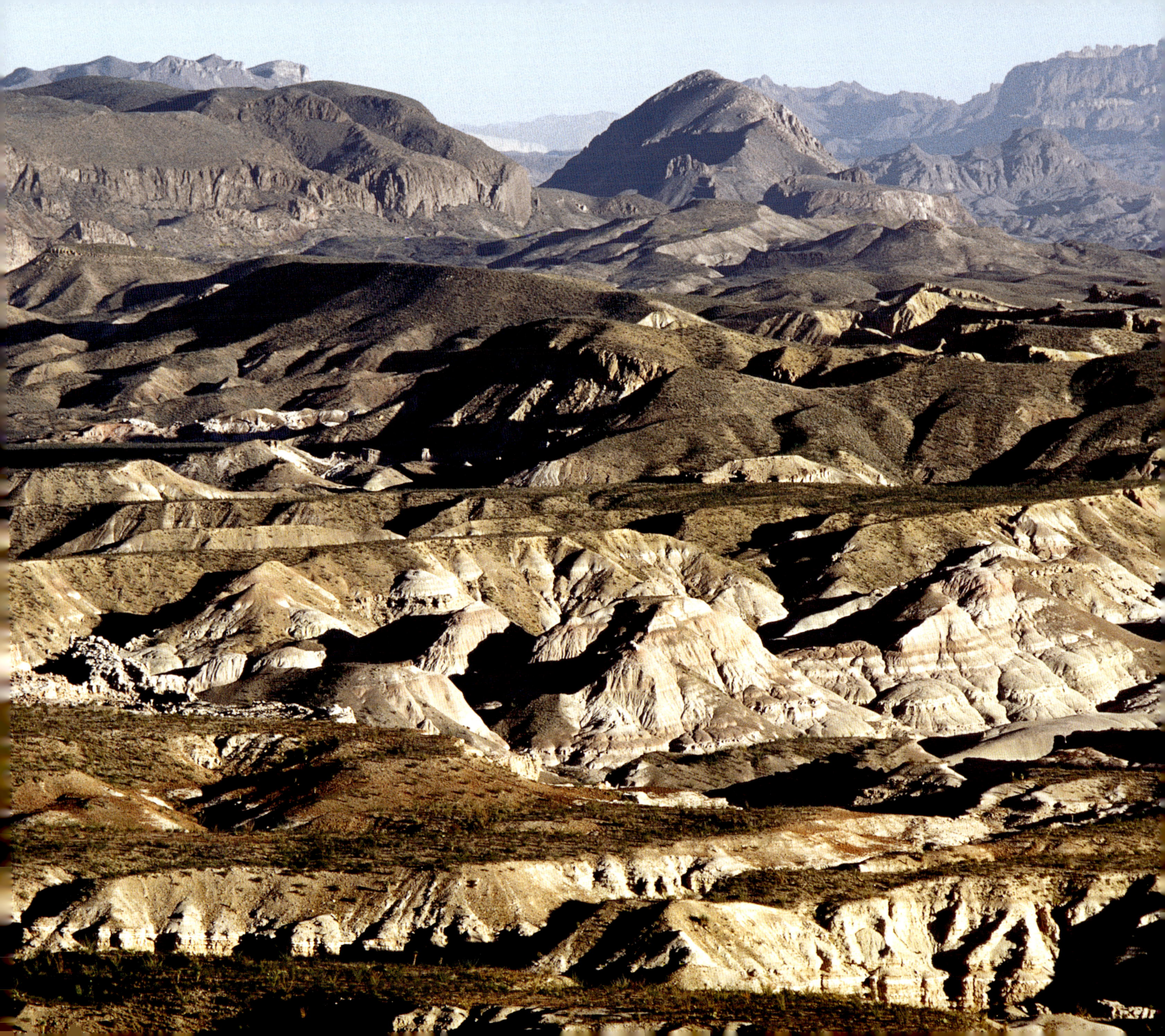

Highly unusual flora and fauna have evolved to live in this strangest of landscapes. The Big Bend lies at the northern fringe of the great Chihuahuan Desert. This is the largest of the four deserts in America, covering 175,000 square miles of Mexico and parts of Arizona, New Mexico, and Texas. While the Chihuahuan Desert might appear empty to the uninitiated eye, it actually contains an astonishing diversity of plant life, approaching that of some rainforests. More than 1,500 different kinds of plants grow in the Big Bend alone. One graduate student studying the Dalquest Site, Karen Killion, discovered forty-five plant species growing in a single 100-meter-square section of seemingly barren desert. More species of birds can be seen in Big Bend National Park than in any other national park in the United States. It hosts an extraordinary variety of insect life, spiders, reptiles, and mammals—including no less than nineteen species of bats.

The natural history of the Dalquest Site, like its geological history, is a single chapter in an epic novel. A surprising diversity of habitats is crammed within its two square miles. In altitude it stretches from 3500 feet at the bottom of Alamo de Cesario Creek to 4340 feet at the summit of Dalquest peak. In between there are springs, riparian zones, tinajas (places where water pools in the rocks), slot canyons, badlands, flood plains, cliffs, alcoves, mesas, caves, creosote bush flats, old lava flows, lava tubes, stands of cottonwood and walnut trees, dense chaparral, and north-facing hillsides carpeted with rare and exotic species of cactus. These habitats have been protected over the years by the site's remoteness and ruggedness. It is a refuge for rare plants that have been driven out of other areas by domestic grazing. Many species of mammals frequent the site, including desert bighorn sheep, bobcats, foxes, raccoons, wild burros, javelina, mule deer, and a host of smaller mammals. Mountain lions often roam the canyons, especially when there are deer around, and the area is rarely free of their kills.

A Parade of Life

WALTER NELSON AND I SPENT A WEEK EXPLORING THE DALQUEST SITE IN LATE summer. A cowboy from the Alazan Ranch (Papa Tom Rawls's old homestead) packed our supplies into the canyon by horseback. There is only one trail into the canyon, and it winds down steep bluffs of volcanic breccia, ash and basalt. Hiking into the canyon is like descending into another world. At the bottom we were overwhelmed with the rich fragrance of wildflowers, desert willows, and cottonwoods. The summer monsoon rains had not yet ended, and a thread of water flowed down the creekbed, playing hide-and-seek among sun-baked boulders. Repeated flash floods had ripped through the canyon that summer, leaving a highwater of debris and tree trunks wedged into rocks twenty feet up the canyon walls. I was reminded of the interesting statistic that the most common cause of death in the desert is drowning.

We pitched camp on a sandy flat above Alamo de Cesario Creek, just inside the red canyon. It was a spectacular campsite. We were surrounded by sheer walls of volcanic tuff, eroded into brilliantly colored curtains and spires—an ancient amphitheater of stone. We made our kitchen and firepit among some boulders of conglomerate and stowed our food in the only shade we could find, under two leaning blocks of stone. We set up our tents on a large apron of soft sand above the curve of the creek.

Opposite page: Chisos Mountains beyond volcanic hills.

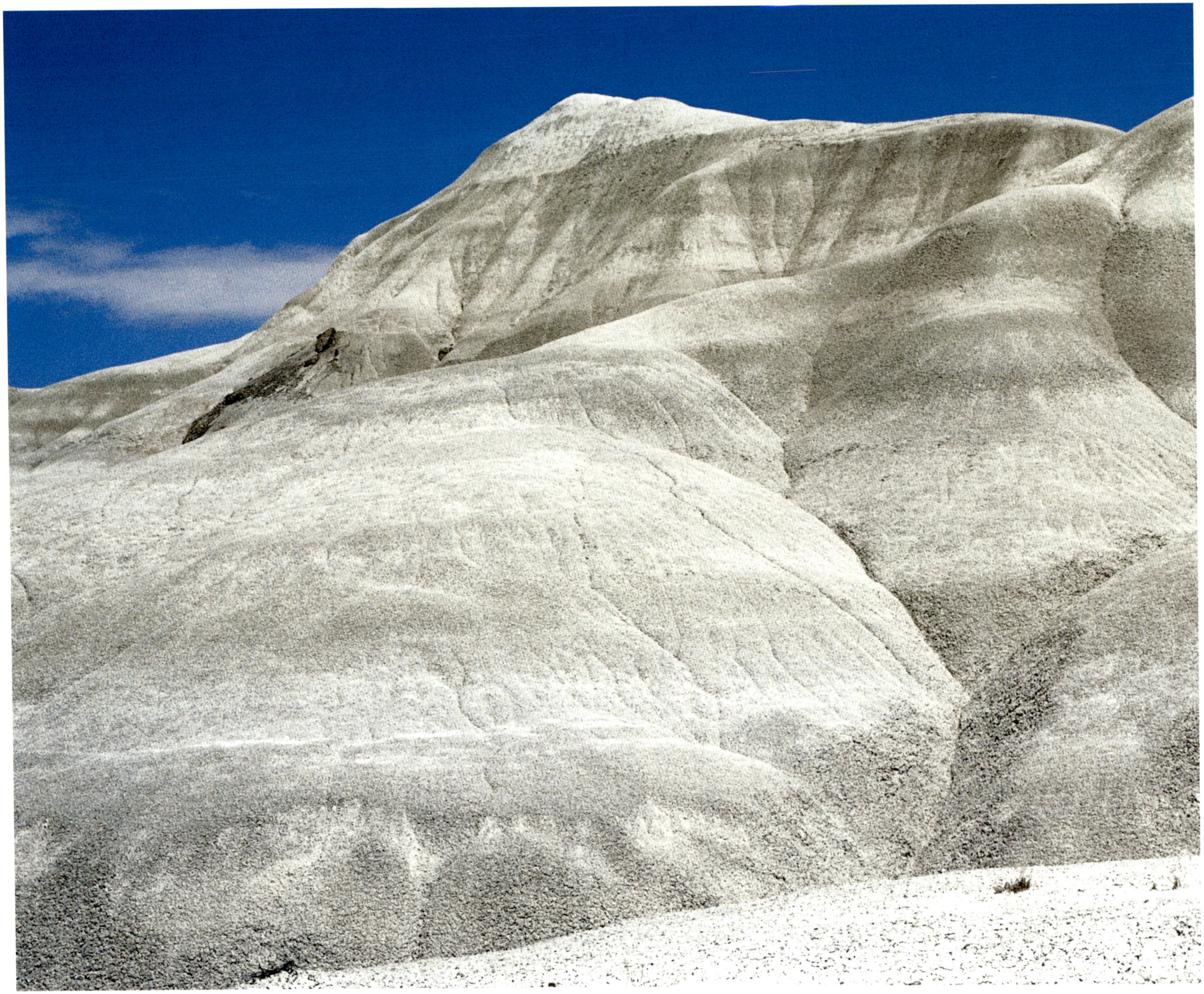

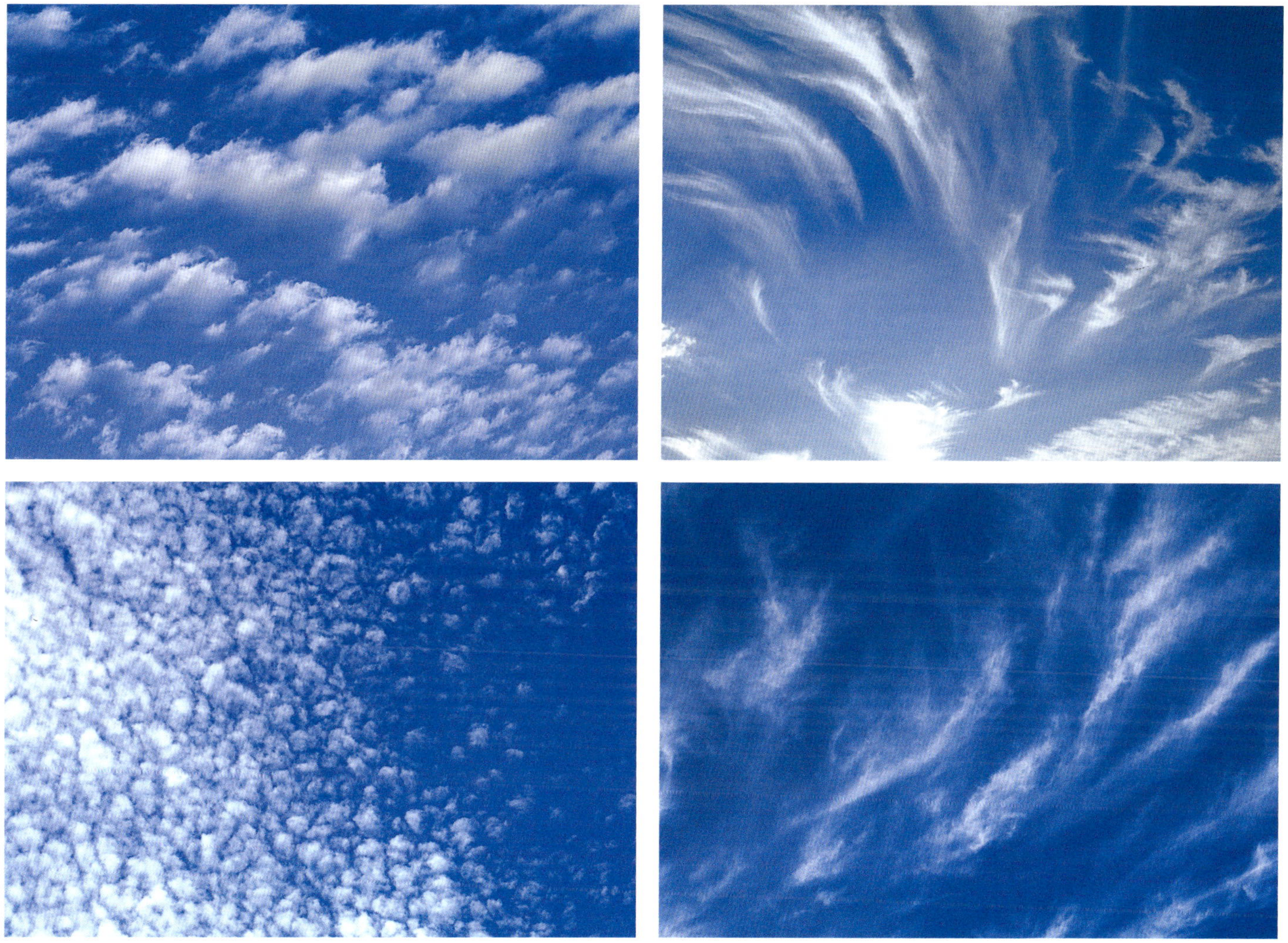

Opposite page: Weathered volcanic hills.
Above: Chihuahuan Desert skies.

We spent the week exploring the canyons, plateaus and badlands that make up the Dalquest Site. Our first stop was Alamo Spring, which lies at the foot of the basalt cliffs formed by the ancient lava flows that once blanketed the site. When it rains, the South Fork of Alamo de Cesario creek forms a waterfall that pours into a bank of cobbles, digging out a pool. The waterfall was dry, but the pool remained filled, fifteen feet long by six feet wide, surrounded by a pebbled beach shaded by desert willows—a sheer paradise.

I set myself down in the shade to watch the parade of life at a desert pool. The sun was directly overhead and it was close to 100 degrees. At this time of day the only animal life foolish enough to be out and about was an array of tough-looking insects and me. Tracks at the pool of mule deer, javelina, a raccoon, a snake, small rodents, and a mountain lion attested to its use by other animals at more respectable times of the day.

The boss of the noonday pool was a large dragonfly perched on a twig above the water. Called a Southwestern Big Red, this is one of the most striking of all dragonflies, blood-red from the tips of its bulbous eyes to the end of its abdomen, with a set of scarlet wings as beautiful as stained-glass. The Big Red feeds by seizing smaller insects in flight. I watched as it unsuccessfully chased one fly after another, each time returning to its twig. Things got interesting when a wicked-looking paper wasp came droning toward the pool. The Big Red was off, swooping down toward it—would there be a fight? At the last moment, realizing its mistake, the Big Red sheered off and looped back to its perch. Meanwhile the wasp landed placidly on the surface of the water, making a dimple but not sinking. These insects have fine setae on their tarsi—hairs on their feet—which allow them to support themselves on the surface tension of the water. The wasp sucked in a large amount of water while the breeze blew it clear across the pool, spinning it around and around in a most undignified manner. Then it flew off, heading back to its nest with a belly full of water. Back at the nest, it would regurgitate the water along with wood and leaf pulp from its stomach and mold it into paper for its nest.

A few moments later another Big Red came flying toward the pool, and mine was off in a flash to defend its territory. An aerial battle took place which would have made Baron von Richtofen's head spin. The intruder was furiously driven off. A moment later my Big Red came back and resettled on the twig, lord of its domain.

The next visitor was a large black wasp known as a tarantula hawk, coming in for a drink. The Big Red rose, flew closer for a look, and wisely returned to its twig. The tarantula hawk is a most elegant-looking wasp, with a glossy black body and deep orange wings. It feeds peacefully on flower nectar, but its mode of reproduction is anything but pleasant. The female cruises the ground until she finds a suitable tarantula spider. She lands on its back, giving it a vicious, paralyzing sting. The wasp digs a hole in the ground, folds in the paralyzed spider, lays her eggs on it, and covers it up. When the larvae hatch they feed on the still-living tarantula, essentially eating it alive until they are ready to pupate. The process takes several weeks and the tarantula doesn't die until the end.

After the tarantula hawk had left, a gold-banded mud dauber wasp landed on the margins of the pool to pick up mud. This black and yellow wasp has a waist as thin and long as a piece of wire, and its nests can be found all over the Dalquest Site, rows of mud cells packed under overhanging ledges of rock. Like the tarantula hawk, the gold-banded mud dauber reproduces in a singularly Darwinian fashion, only to be preyed upon

Opposite page: Oligocene tuffaceous mudstone.

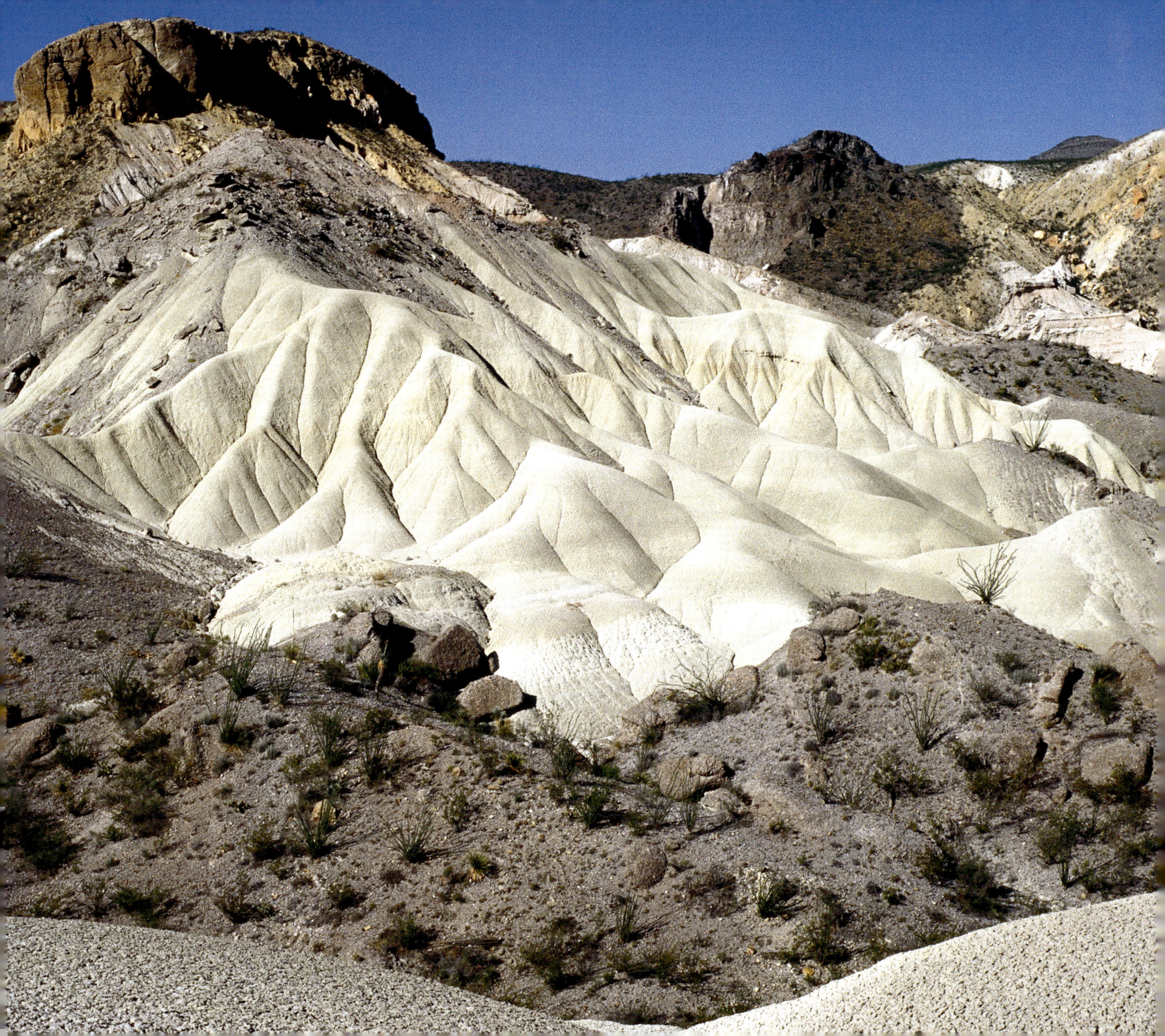

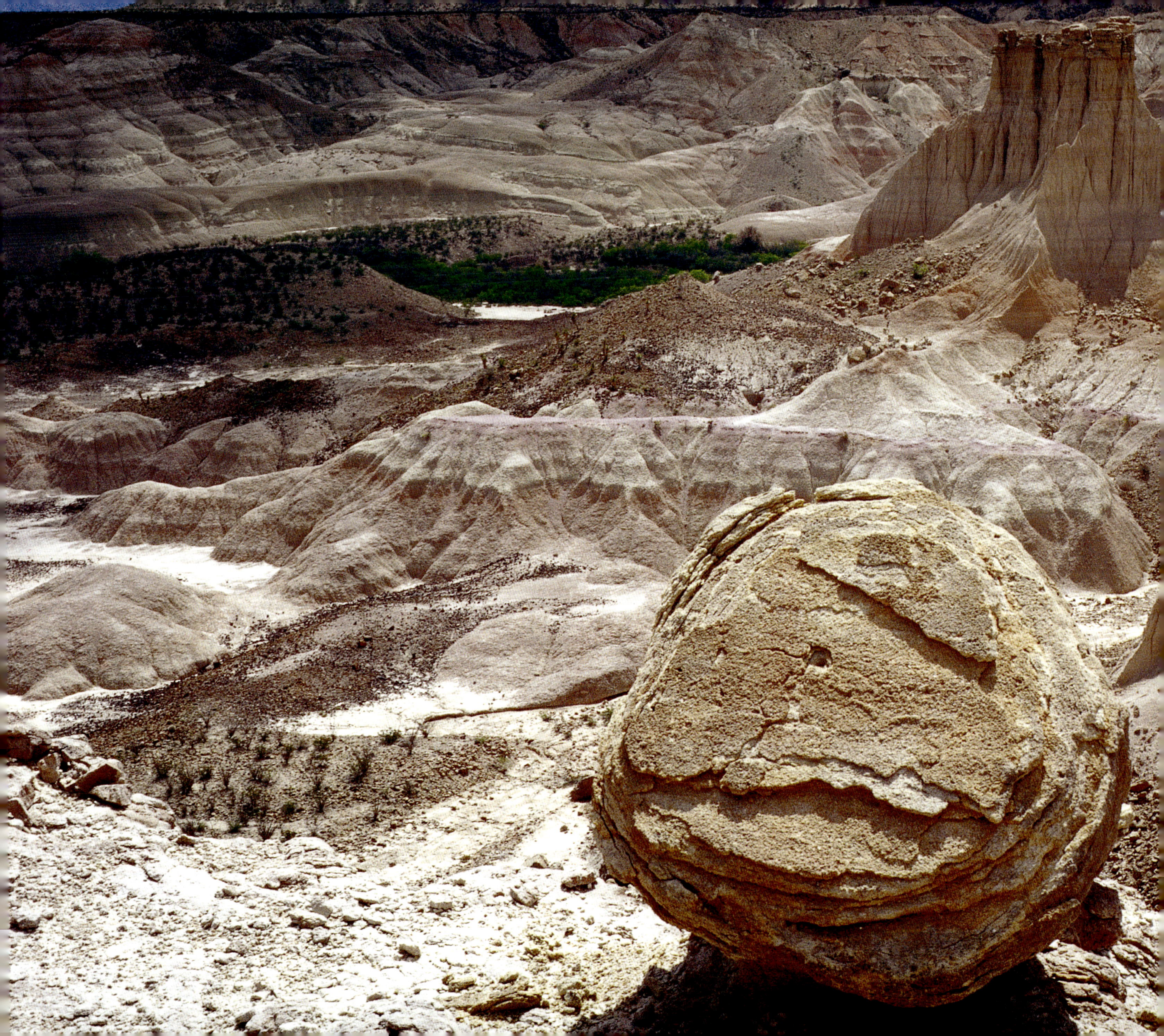

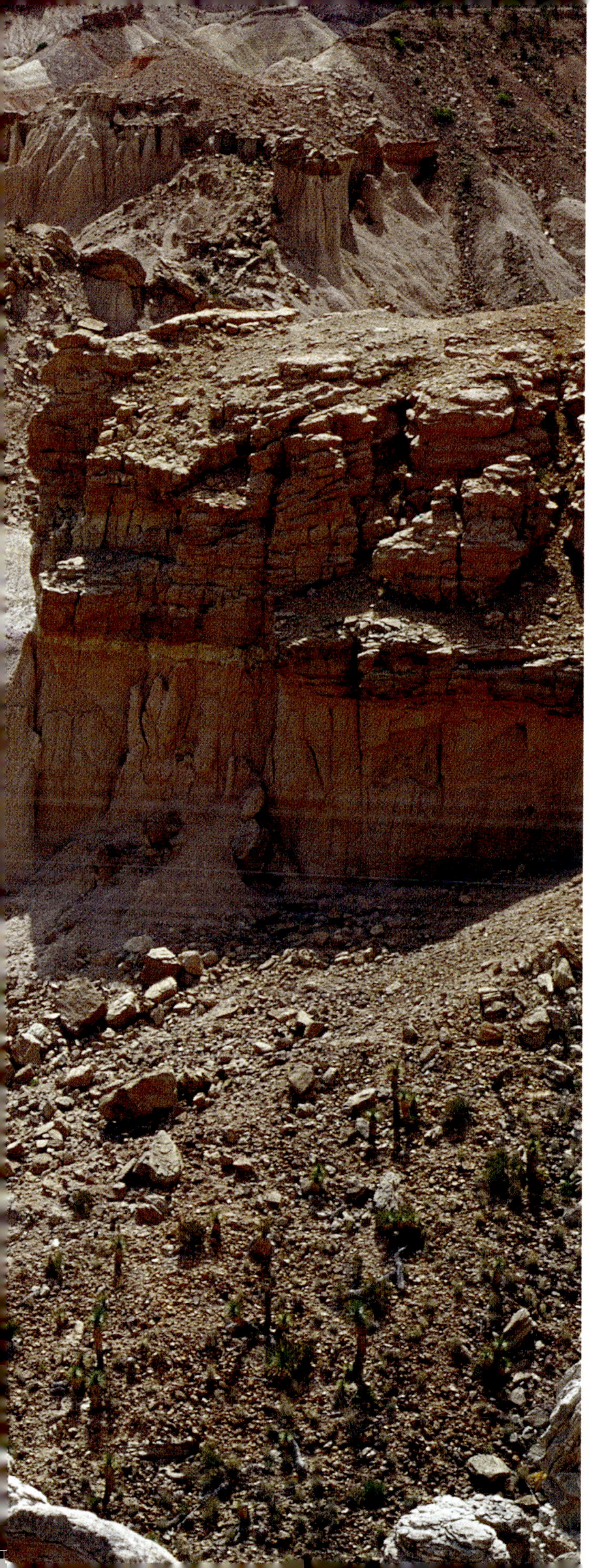

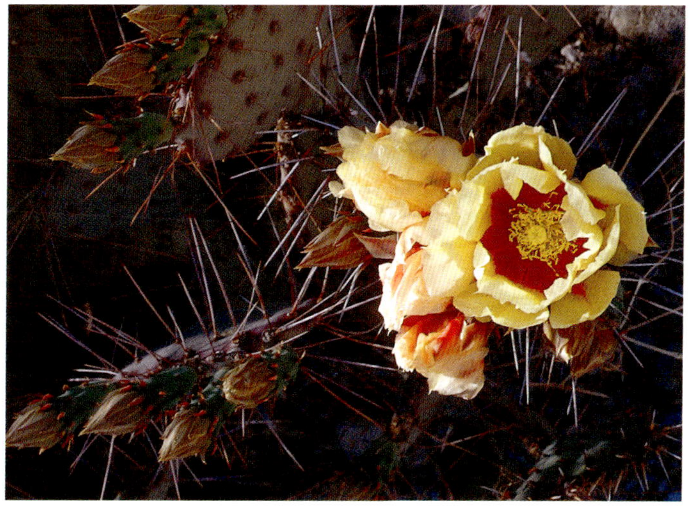

Opposite page: Black rock basin, looking east.
This page: Flowering prickly pear.

itself by its even more evil cousin, the blue mud dauber. The gold-banded mud dauber builds a series of mud cells and lays an egg in the back of each one. It packs each cell with twenty to thirty small, paralyzed spiders, then caps it off with mud. When the egg hatches, the larva will eat its way through this paralyzed (but living) food supply before pupating.

The blue mud dauber doesn't bother to build nests of its own. It searches out the cells of its cousin, which it tears open, dumping the spiders and eggs to the ground. It then collects its own paralyzed spiders (it favors black widows), packs them into the cell, lays its own egg, and closes it up. Entomologists were first alerted to this strange behavior by puzzling over why there were little piles of dead spiders underneath certain mud dauber nests.

As I waited at the spring, more wasps came and went, along with a steady stream of other insects, all enjoying the luxury of a pool of water in the desert. Behind me thousands of honeybees droned in the blooming desert willows, gathering nectar and pollen. Water striders flitted about the surface while water beetles, carrying their own air supply in the form of a bubble underneath them, plied the mahogany-colored depths of the pool. I could see dragonfly naiads crawling along the bottom, hunting for food. When I left Alamo Spring the Big Red was still perched on its twig, waiting in ambush, king of the pool.

Walter and I hiked down to the second source of permanent water in the canyon, Fern Spring. The water seeps from a fissure in the rock, creating a wall of cascading maidenhair ferns. Just below the spring is a natural campsite, hidden among willows near the mouth of the canyon. As we continued on, I found an Indian scraper lying below this campsite, flaked out of a fine, mustard-yellow opaline. Throughout the canyon I found examples of flakes and scrapers made from this same opaline, the last traces of the area's long-vanished Indian inhabitants. On the last day I discovered the source of the material—on one of the basalt flows high above Alamo Spring. The lava had hardened with many bubbles in it, and these cavities later became filled with calcium crystals, agate, or opaline. In one place nodules of yellow opaline were eroding out of the hillside, and all around lay chunks and flakes knocked off by the Indians while they collected the material.

Another day Walter and I hiked up to the head of the white canyon, the middle of the three canyons that come together to form the Dalquest Site. The upper part of the canyon is accessible only by a fixed climbing rope, which Walter had previously rigged. The canyon was almost entirely white, creating an eerie winter-like wonderland broiling in the 100-degree-heat. Everywhere we could see evidence of the violence of recent flash floods: the canyon bottom scoured and dug out, riparian plants pressed flat, huge boulders recently moved. We climbed over two more pourovers before arriving at the head of the canyon. It ended abruptly in a great stone alcove, its roof nubbled with the mud nests of swallows. When we arrived, a flock of doves burst out of a pool, their wings rustling like silk as they passed overhead. We explored several side canyons before finding a way up and out of the canyon to the rim.

Opposite page: Volcanic hills behind conglomerate boulders.

From there we climbed straight up the face of Dalquest peak, which rises above the west rim of the white canyon. We scrambled up a steeply pitched slope of volcanic rock, dotted with spectacular cacti and succulents,

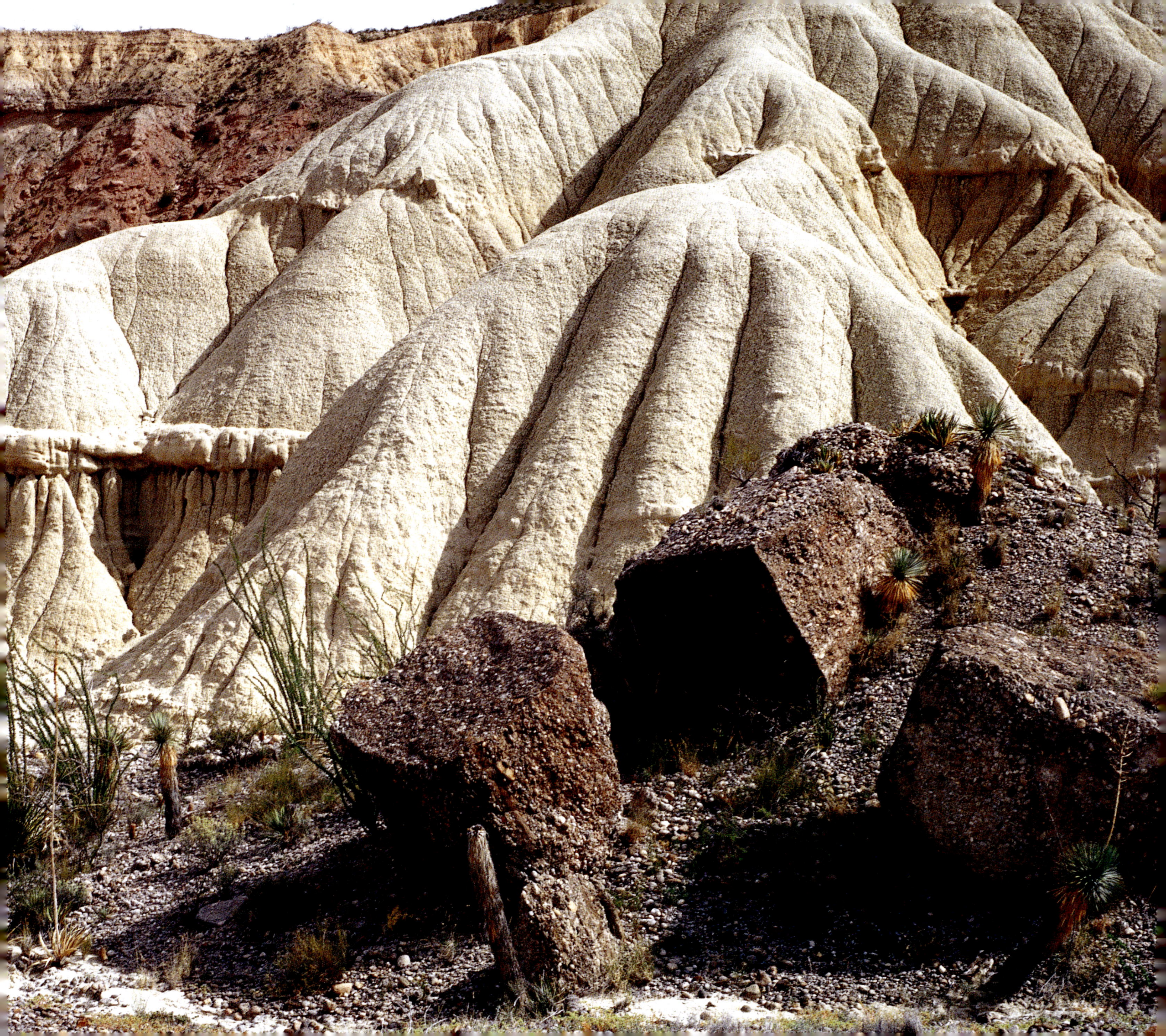

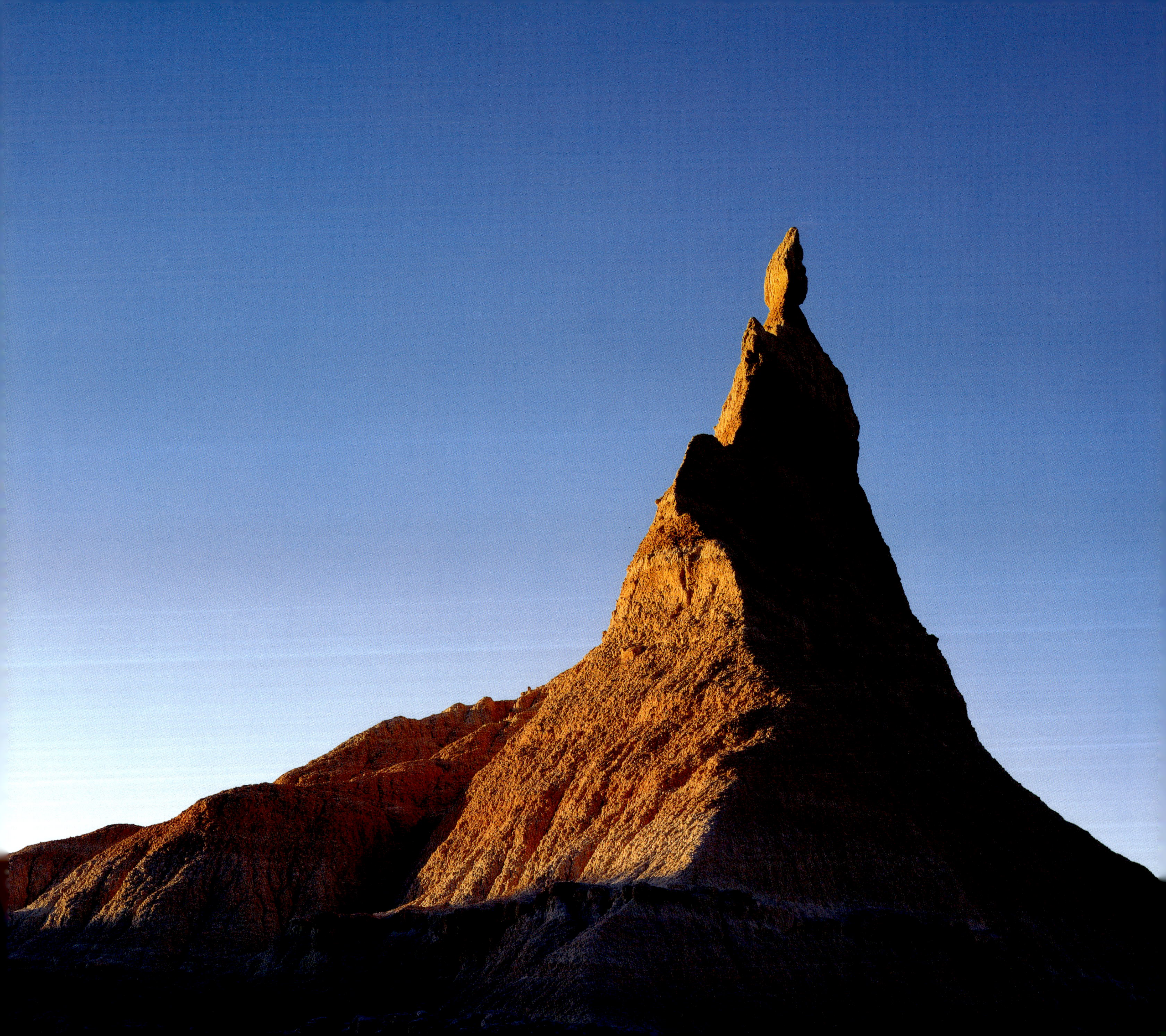

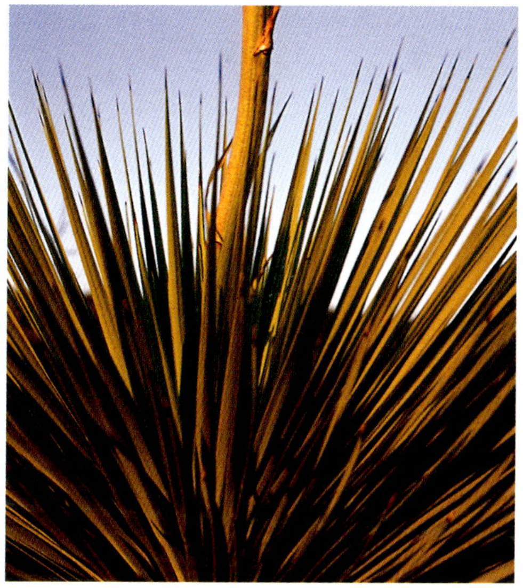 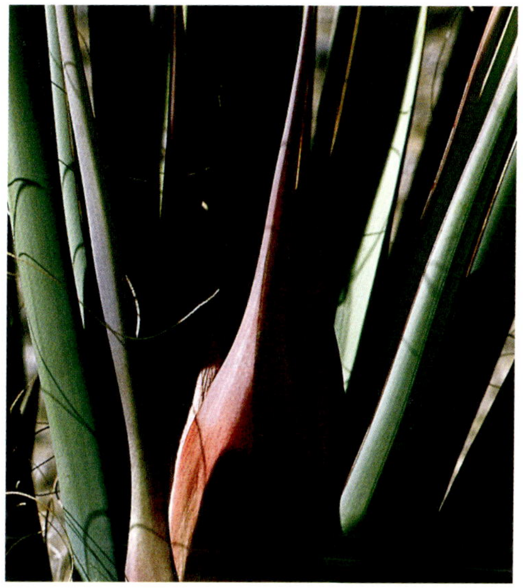

Opposite page: Giant wedge at sunrise.
This page, left: Sotol. Right: Yucca leaves.

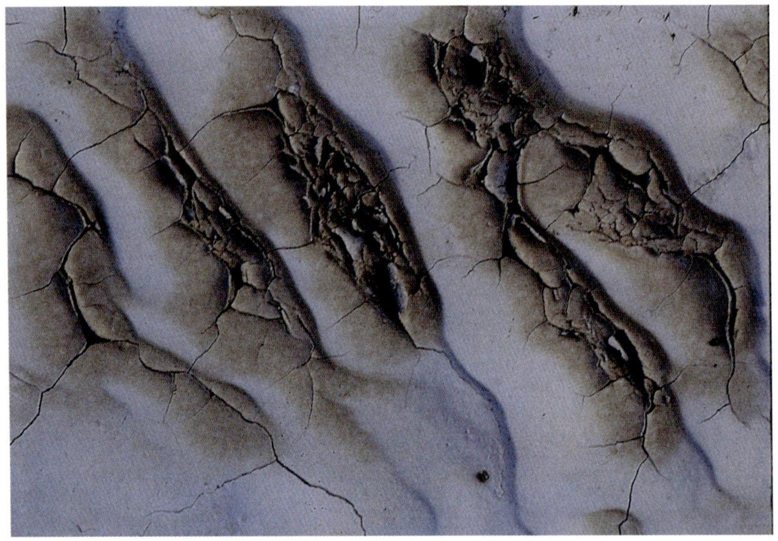

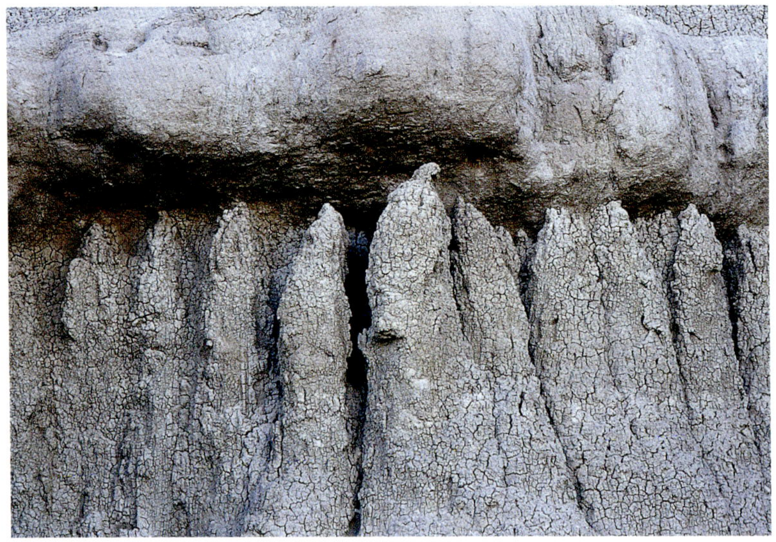

This page, above: Mud cracks on ripple marks in a stream bed.
Below: Water-eroded formation.
Opposite page: Sunrise, Dalquest peak beyond rim.

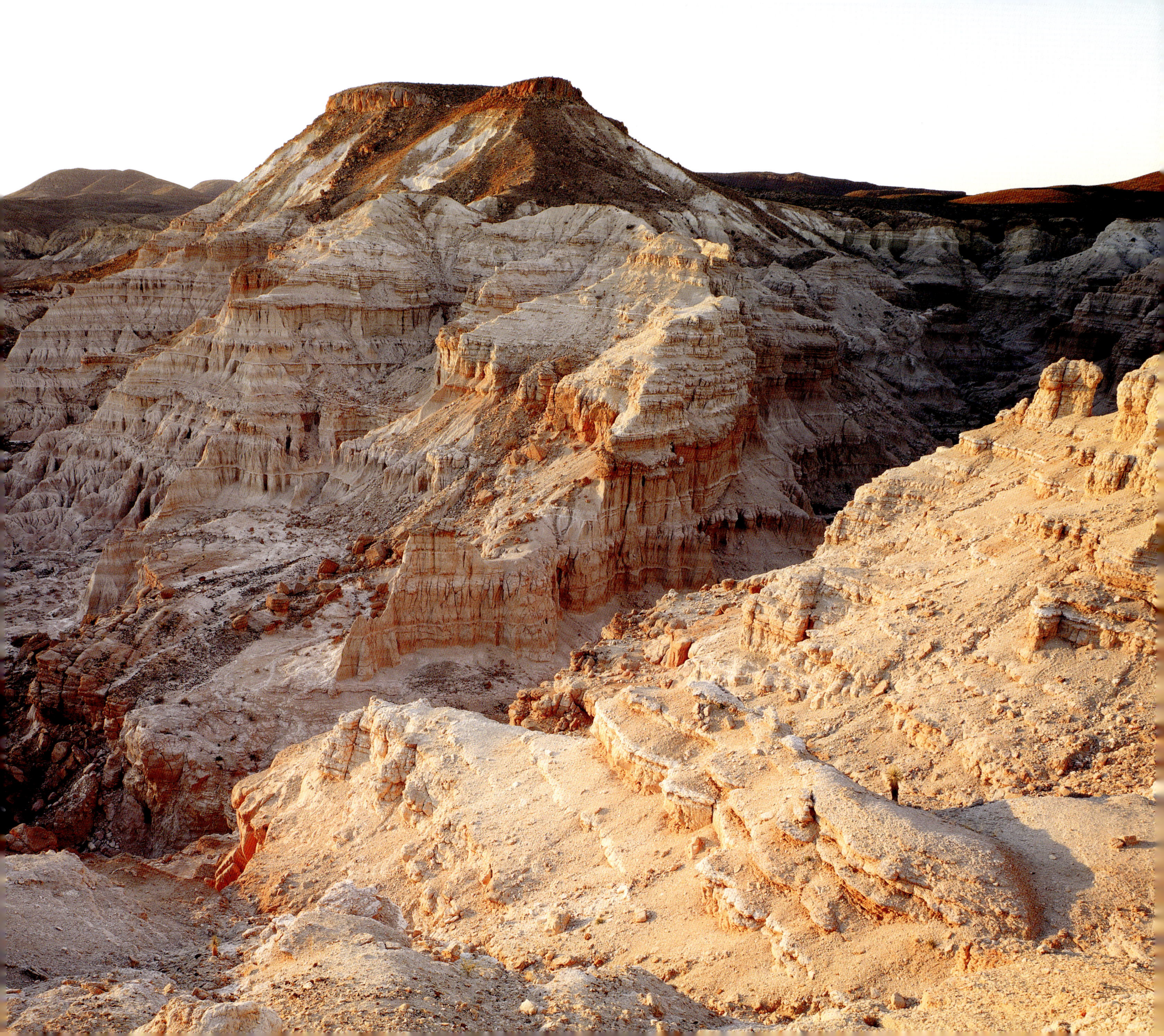

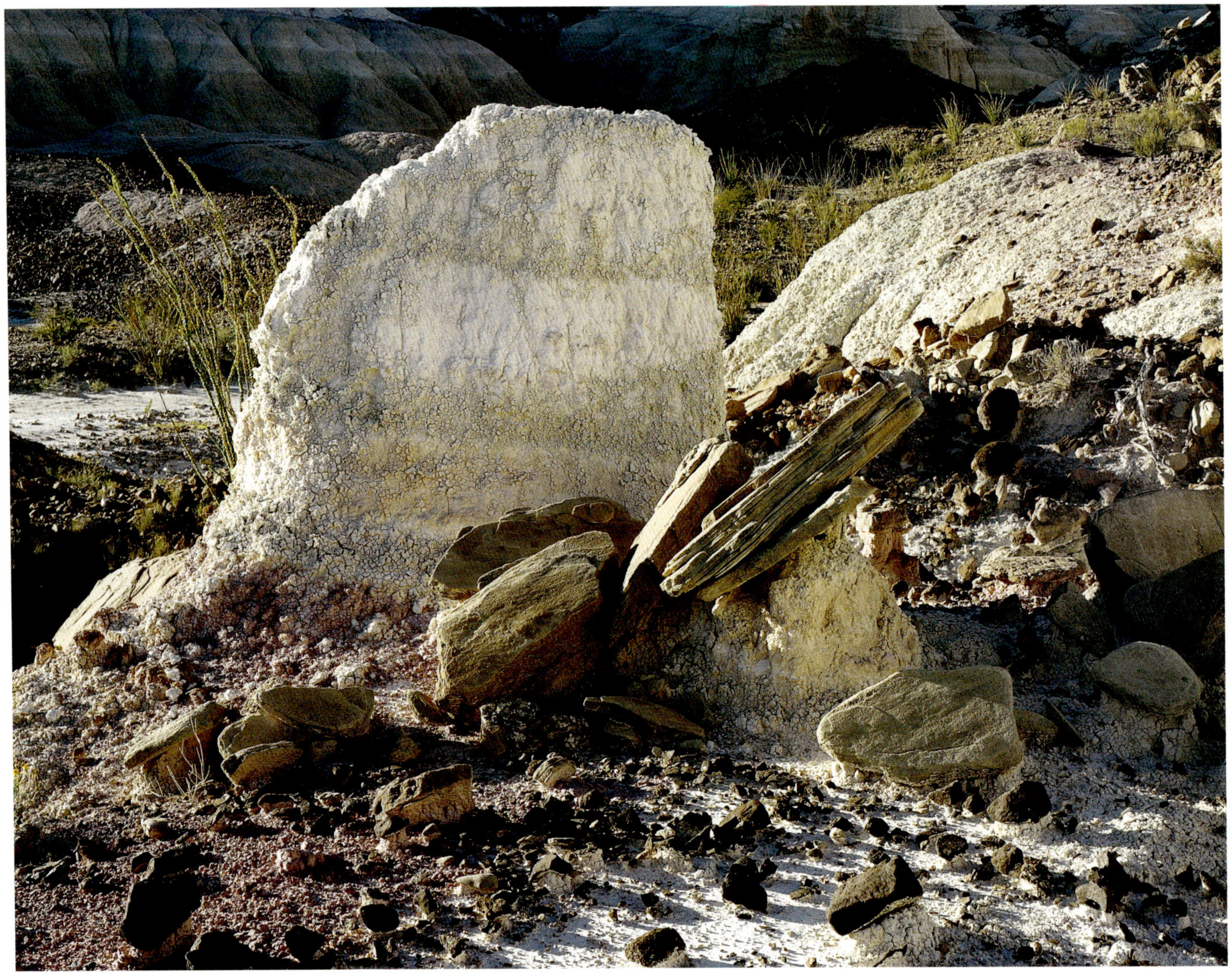

This page, and opposite: Formations in black rock basin.

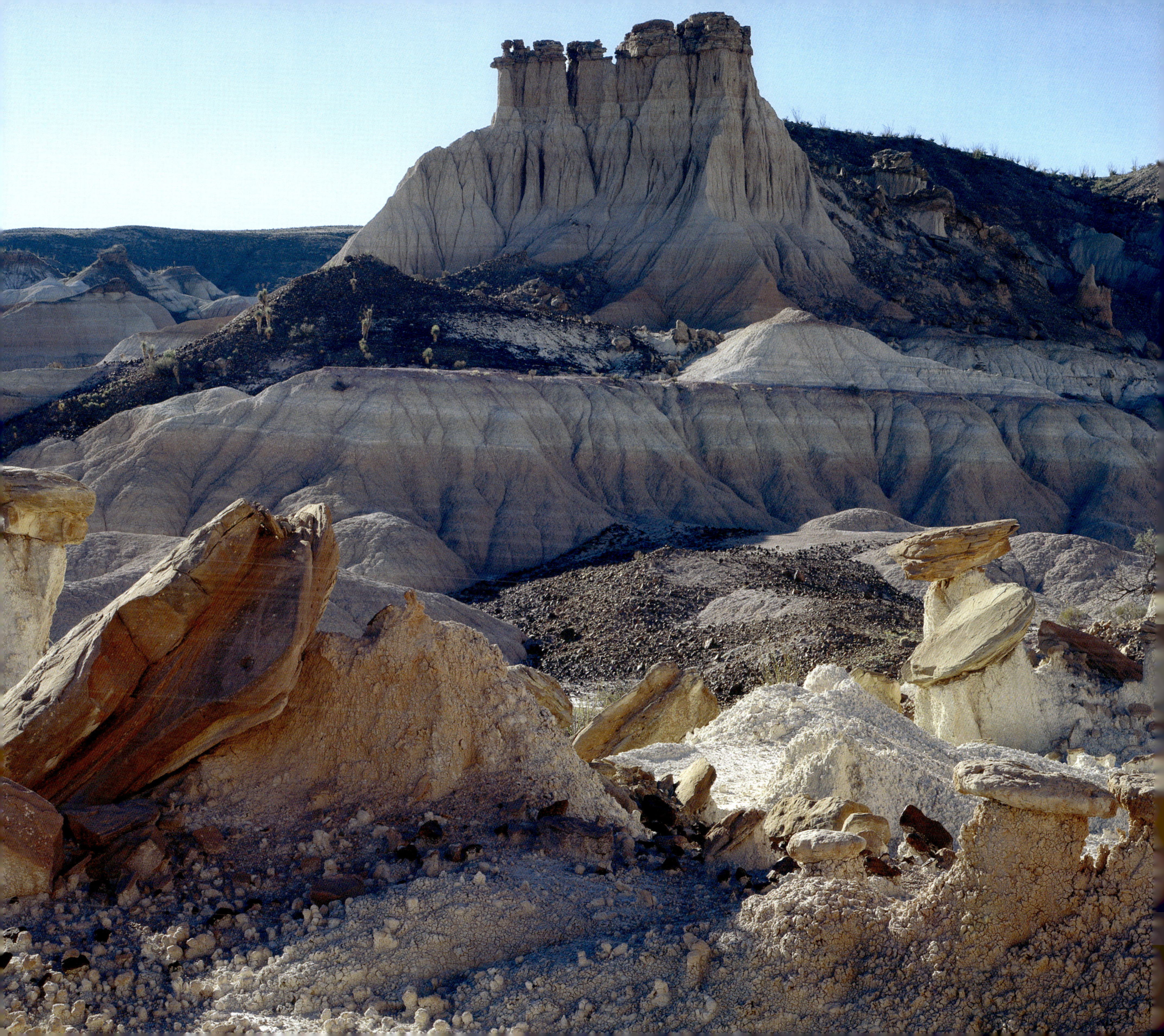

many in breathtaking bloom. One early traveler in this area complained that "Each plant in this land is a porcupine. It is nature armed to the teeth." And so it seems until a close inspection of the ground revealed that at least some plants survive through subterfuge. I found clumps of resurrection plants, so called because they appear utterly lifeless, black, and shriveled when it is dry—only to uncurl and burst into vibrant green life at the first rain. The ground harbored a dense population of living-rock cactus, a low-growing succulent that is almost impossible to see when not in bloom. These, however, were in bloom, each one unfolding a single, stunningly perfect pink and yellow flower from its center. The creosote bush employs a different survival strategy: its leaves are loaded with unpleasant chemicals, making them impossible to eat, and to discourage plant competition it actually poisons the very ground around it. There were, naturally, plenty of plants "armed to the teeth" with spikes, spines, thorns, needles, and a fuzz that penetrates the lips and tongue of any animal foolish enough to take a bite. I identified a number of unusual cacti. These included the horse-crippler, an aptly named cactus that grows almost flush with the surface, exposing only a cluster of spines as thick as ten-penny nails and infinitely sharper. Also growing on this single hillside were devil's head, cane cholla, purple prickly pear, Christmas cactus, button cactus, Turk's head, and sotol. Parts of the hillside were carpeted in Texas false-agave (not an agave at all but related to the pineapple) and lechuguilla, the "indicator" plant of the Chihuahua Desert.

The slope ended at the summit of Dalquest peak, commanding a spectacular view of Big Bend National Park, Big Bend Ranch State Park, and the mountains of Mexico beyond. The mountains seem to go on forever, like a sea frozen in violent agitation, the peaks shading from purple to light blue as if reaching up to become part of the sky. We were not the first to enjoy this view. Artifacts and flakes have been found on the peak, indicating that in days gone by bands of Indians had stopped here to flake out projectile points, while watching for game or travelers on the two trails that passed on each side of the area.

We climbed back down into the canyon along a steep slope of volcanic ash, to reach an area called the wedge plateau, named for a blade-like rock formation resembling the giant fin of a shark. The wedge plateau is covered with bizarre geological formations, including a squat arch that looks like the Arc de Triomphe in Paris. Surrounding this arch is a field of black rocks resting on a bed of volcanic ash as white as snow, forming a striking visual impression. Nearby, dozens of rocks stand on pillars of the same ash, some ten to fifteen feet high.

While Walter photographed, I sought a place where I could write in my journal. I finally found a pool of shade cast by a tall pillar of ash on which balanced a rock the size of a car. There I made a curious discovery: scattered about were a number of chips and flakes of stone. An Indian had stopped here just as I had, probably to rest in the same shade, and had flaked out a stone tool before going on his way, leaving the flakes behind as prehistoric trash. There they had rested ever since, for so long, in fact, that some were balancing on their own half-inch-high pedestals of ash. This gave me an idea, and I did a rough and probably grossly inaccurate calculation in my notebook. If these flakes had been left, say, during Victorio's war one hundred and fifty years ago, that would give an approximate rate of erosion of the ash at about one inch every three hundred years. The bal-

Opposite page: Volcanic ash interbedded with sandstone.

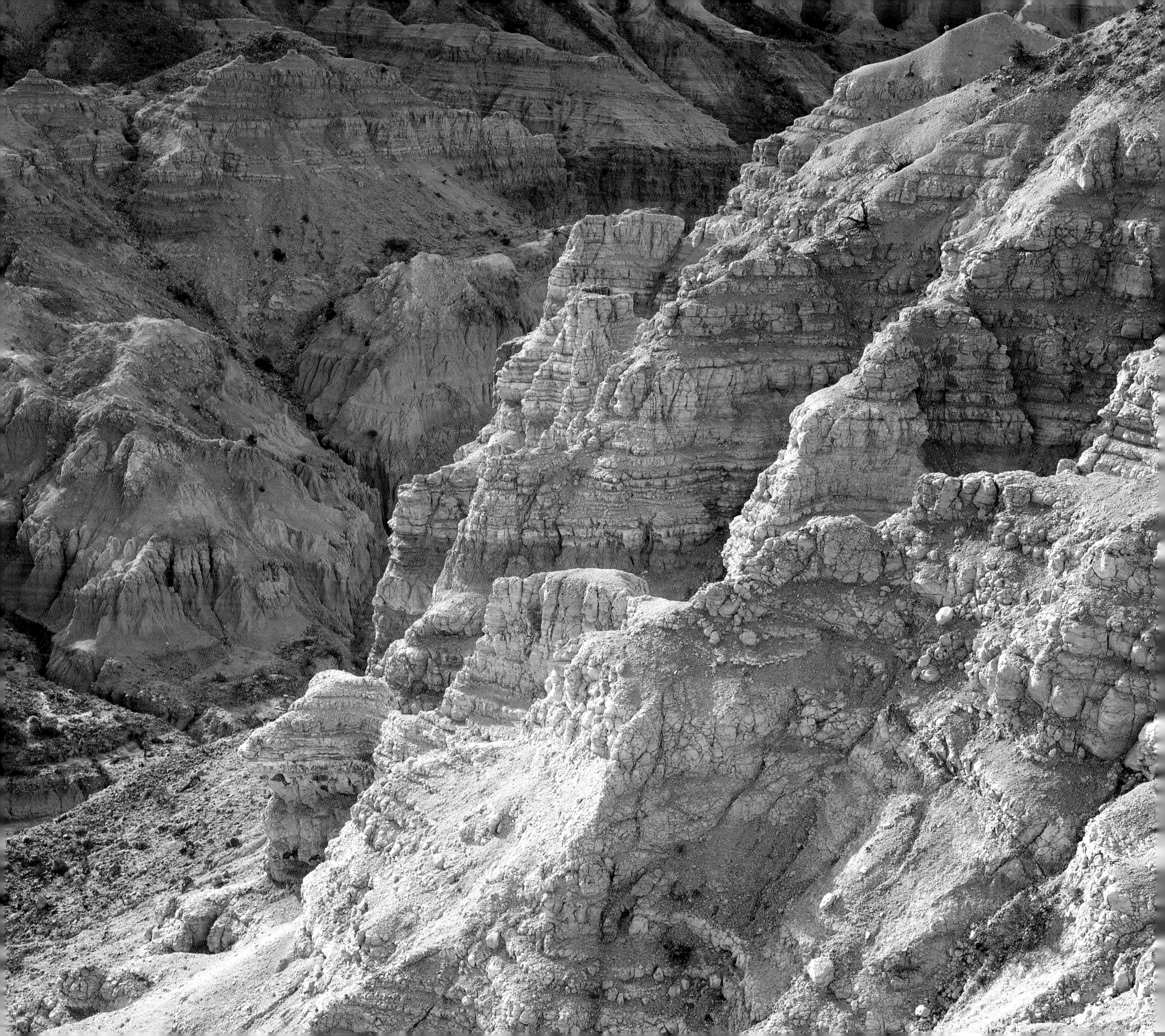

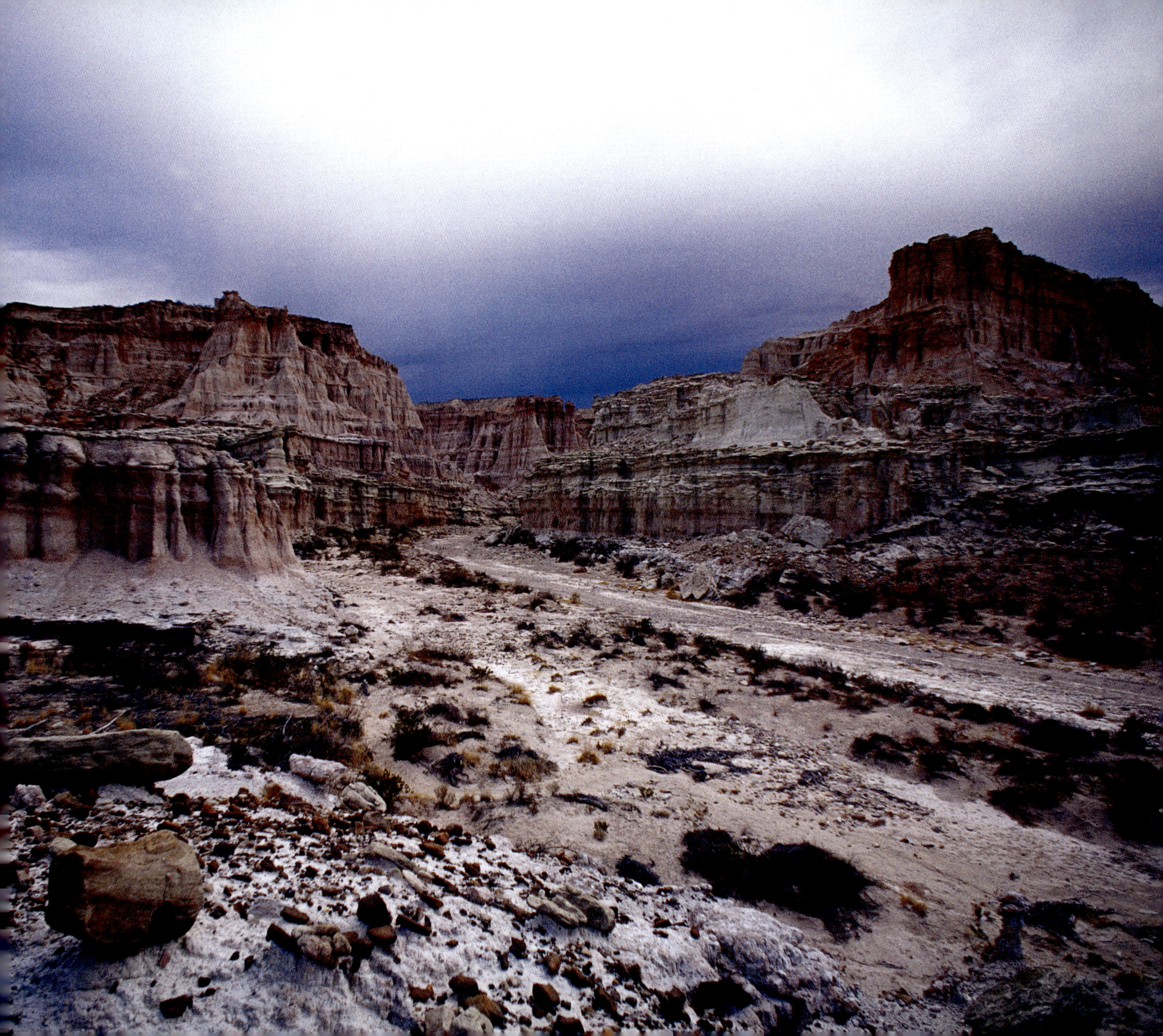

ancing rock stood on a pillar of ash that was seven feet high—which meant the rock hadn't moved for at least twenty-five thousand years.

The red canyon, which lies partly in the Dalquest Site, is the most spectacular of all the canyons in the area. Here the Great Creator ran riot, carving the layers of colored ash into a stupendous variety of shapes—laying bare the very ribbons of time. The word "red" hardly describes the infinite hues of vermilion, cinnabar, pink, coral, peach, blush, chocolate, and cream which tint the layers of this canyon. The beds of tuff alternate with harder layers of sandstone, which were once the floors of ancient lakes. As we explored farther up the canyon, each bend disclosed a fresh wonder. In one place, the walls formed five-hundred-foot-long draperies of stone. Just beyond that a striated dome rose up, bearing an uncanny resemblance to the famous Duomo in Florence, Italy. Rising on all sides was a bewildering array of rock formations—buttresses, arches, fins, balanced rocks, spires, ribs, crevasses, alcoves, and caves.

A thick band of conglomerate runs through these layers. This interesting formation was once the bed of the ancient river mentioned earlier, which had its headwaters in the nascent Rocky Mountains and flowed eastward through here on its way to an ancient sea. The old riverbed had hardened into a band of rock much stronger than the surrounding ash. Erosion was now undermining this layer, and pieces of it the size of small houses had broken off and tumbled into the wash. There, the tremendous force of flash floods was pushing them downstream, piling them up in jams. Some of these boulders had split along the deposition layers, exposing fifty million-year-old ripple marks made by river currents.

A stream of muddy water flowed down the red canyon, forming pools large enough to swim in, and several waterfalls. After taking a shower alfresco, I rested in the shade of an overhanging cliff, next to a small catclaw bush. A Scott's oriole came to rest on a branch not twelve inches from me, peered this way and that with beady black eyes, and then flew off, having left a small dropping on the ground. Five minutes later a giant millipede came rolling along. These are the most dramatic arthropods in the canyon, up to five inches in length, tubular, brown and glossy. They go abroad in the noonday sun when the temperature at the ground approaches 150 degrees. Although they are harmless, when disturbed they curl and twitch grotesquely, exuding a foul, brown substance. I watched as the millipede stopped at the dropping, ate it, and continued on its way. In the desert, nothing is wasted.

Rare mammalian life abounds in these canyons. Mountain lion tracks pad the sandbanks, mingling with the crisscrossing tracks of wild burros and deer. Bighorn sheep trails web the badlands and bluffs, and we often came across the sandy, scooped out areas where they bed down. For several days we were circled by a pair of hopeful vultures, which no doubt wondered when the two slow, sickly animals hobbling along below them were finally going to drop. (We were glad to disappoint them.) We never did see a mountain lion, but we saw plenty of burros, deer, and javelinas—and one magnificent desert bighorn, three and a half feet high at the shoulder, with a curving rack several feet in length. It watched us, wary but not alarmed, before ascending a seemingly unscalable cliff and disappearing.

Opposite page: Carved buttes at the entrance to red canyon.

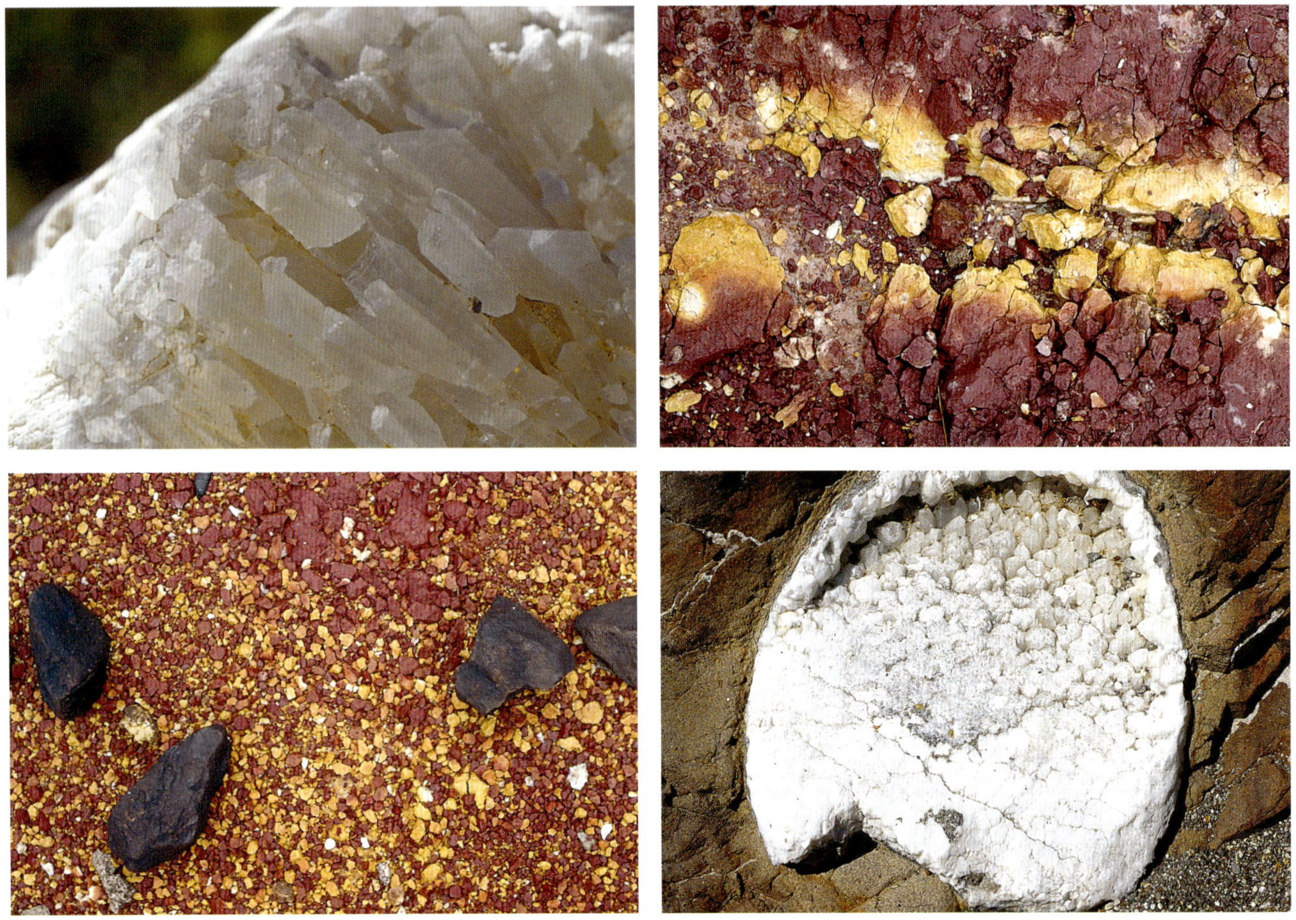

This page: Calcite crystals in basalt.

Opposite page: Pitaya cactus.

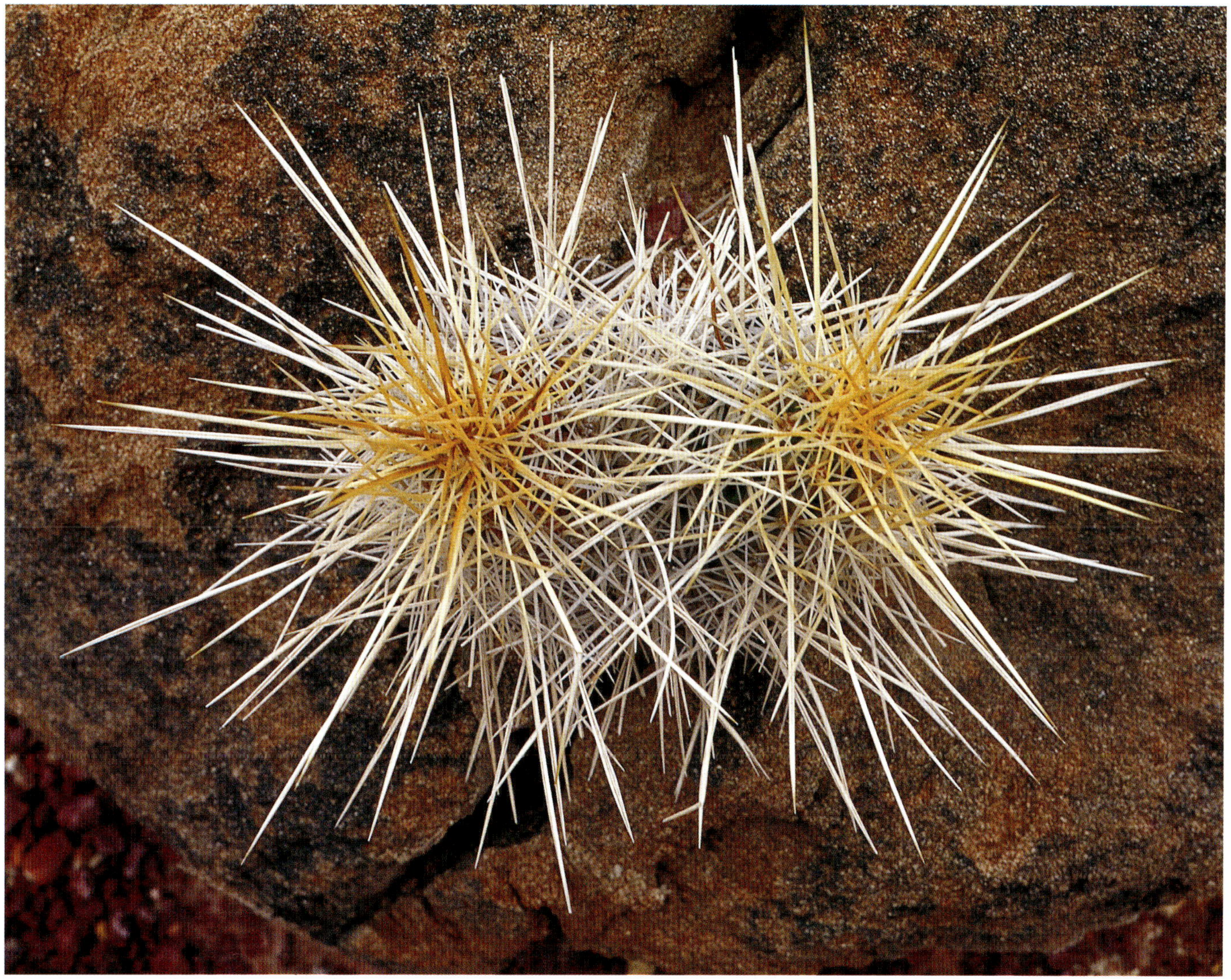

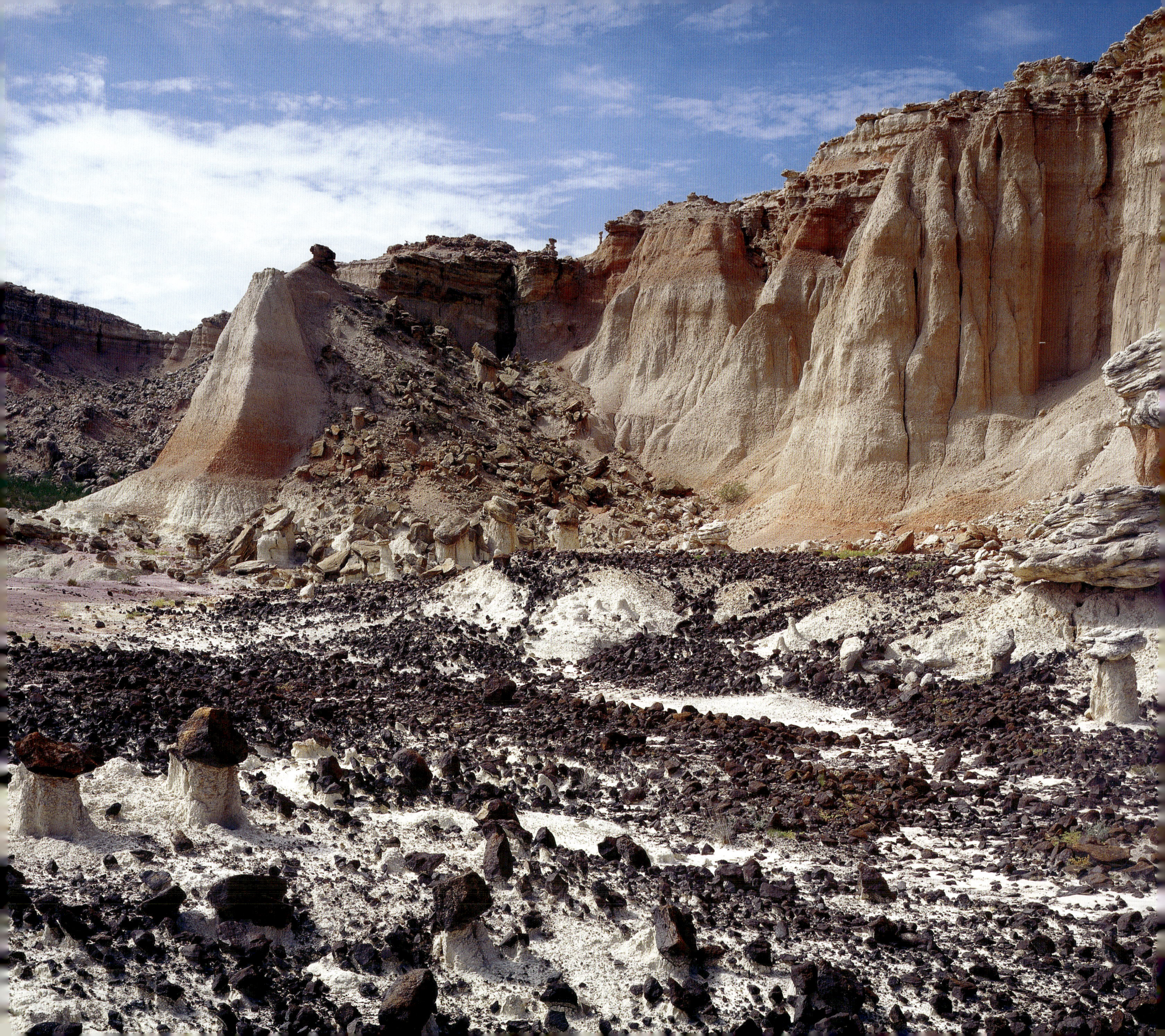

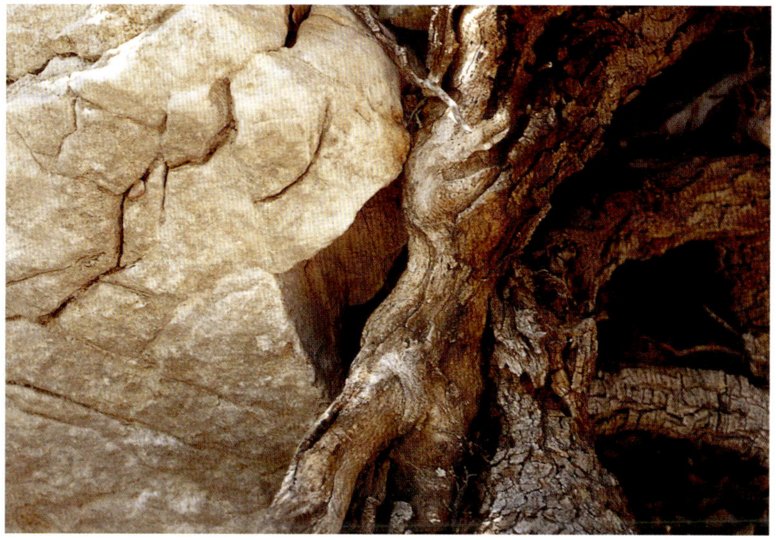

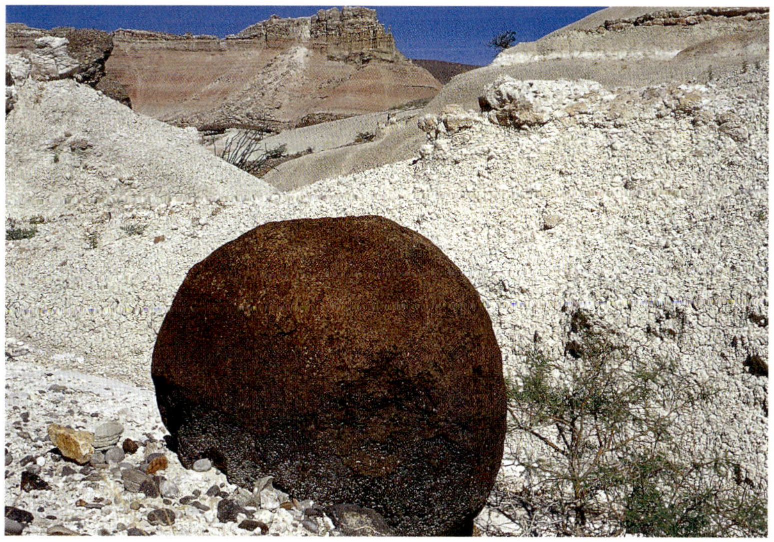

Opposite page: Purple earth in Devil's Graveyard.
This page, above: Rock-breaking oak root.
Below: Lone basalt boulder in volcanic hills.

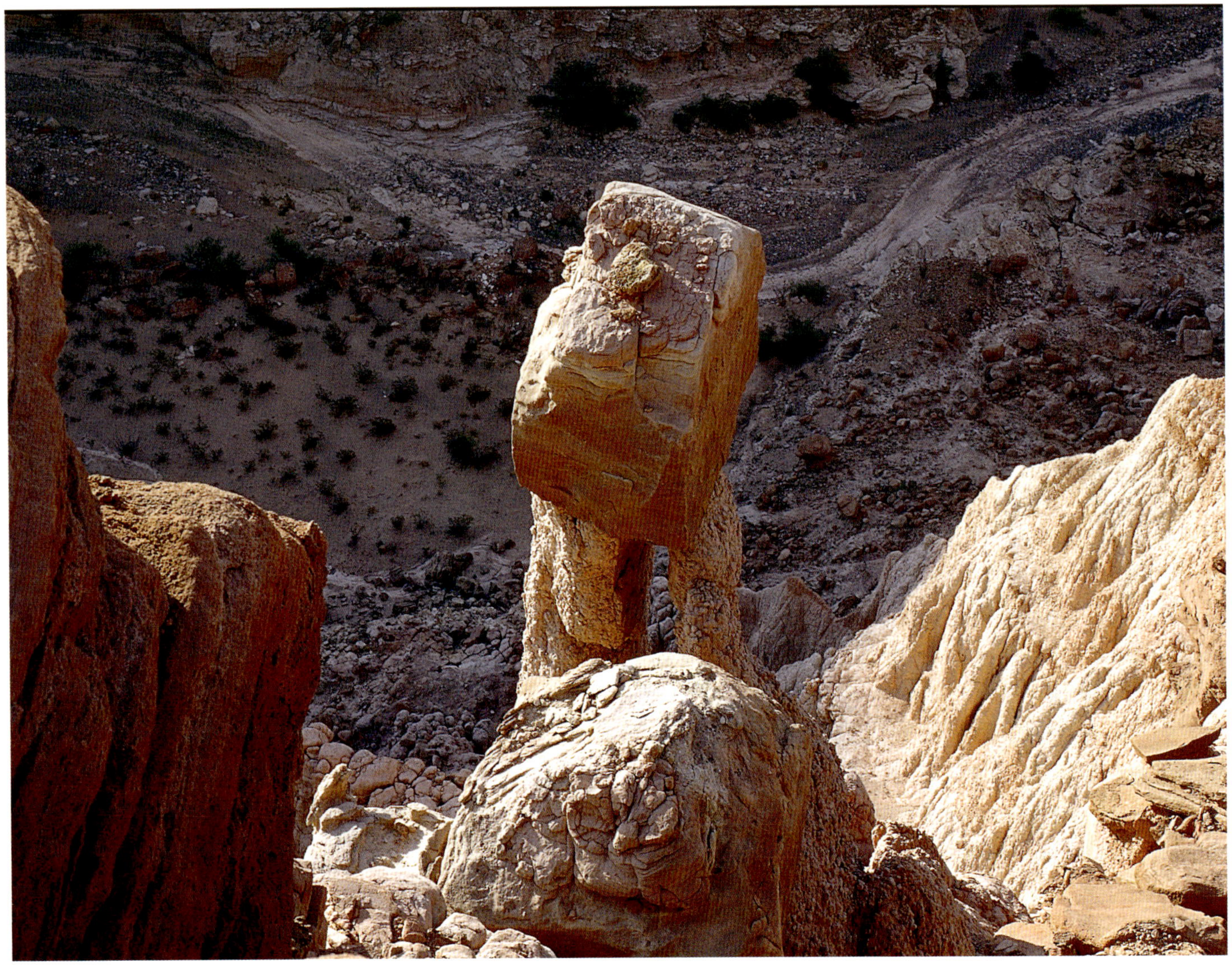

This page: Hoodoo.
Opposite page: White canyon.

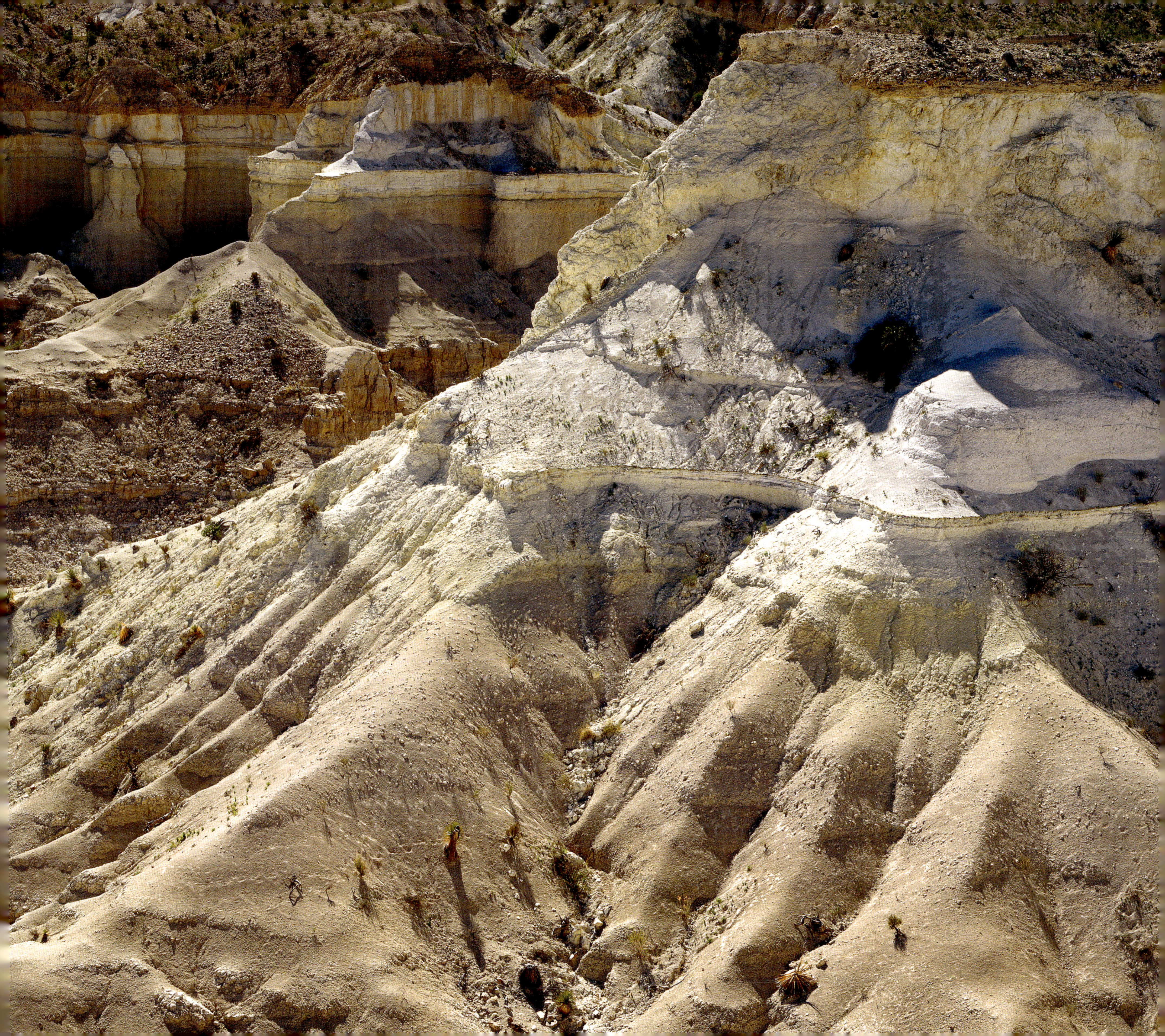

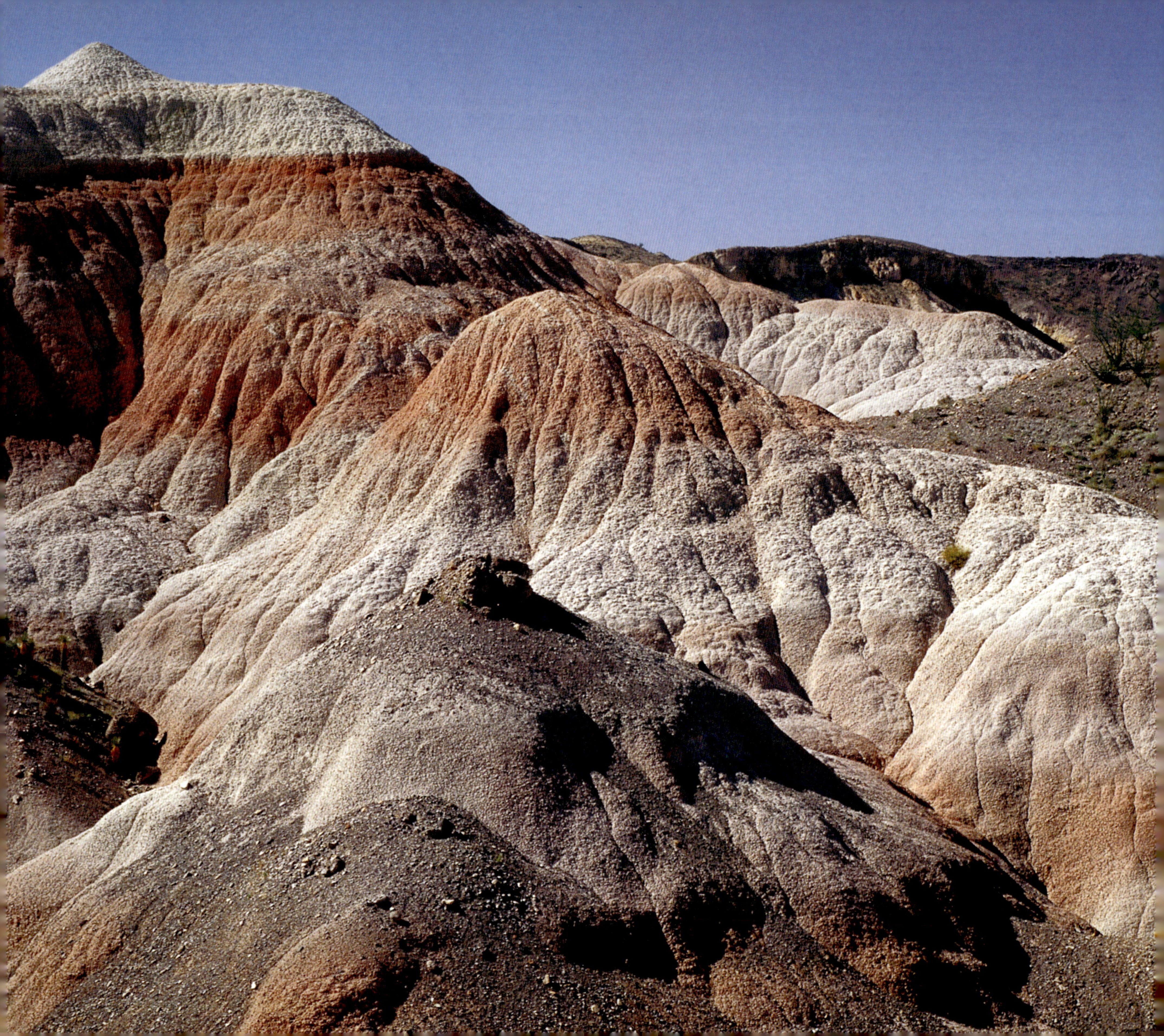

A Work in Progress

IT WAS STILL THE TAIL END OF THE RAINY SEASON. EVERY DAY THUNDER-clouds would boil up over the Sierra Madre mountains of Mexico, spread their anvil heads against the tropopause and creep our way. They would be gone the next morning, only to reappear in the afternoon, bigger and more threatening. One evening the storm finally broke. By sunset the sky was covered with roiling clouds, and thunder boomed through the canyons. Heavy drops came spinning out of the sky, sending up puffs of sand, and soon there came a violent downpour. It stopped fifteen minutes later, as suddenly as if a spigot had been turned off. We could hear the gurgle of flowing water coming from all directions. We exited our tents to find hundreds of streamlets pouring off the canyon walls and rushing madly into the creek. Each stream was a different color depending on what sediments it had encountered—from milk-white to dark chocolate. The creek itself had surged into a swift mass of muddy water six feet wide and two deep. I stood outside the tent, enthralled at the sight and sound of so much water, inhaling the rich perfume that the rainfall had released from the desert.

Nature had not finished with us. As night closed down, a much bigger storm rolled in. The sky grew black and the canyons moaned with gusts of wind. Lightning flickered continuously beyond the canyon rims, getting ever closer, the rumbles coming in like a barrage of distant artillery. Soon we could see stabs, forks, and fingers of lightning splitting the sky and striking the rims all around us, a bolt every few seconds, the crashes echoing and rumbling away down the canyons. One bolt reached the bottom and struck near our camp, a flash-boom that split the air. A few minutes later, just when it seemed the storm was passing, we heard an unfamiliar noise. It started like a herd of stampeding horses, then it swiftly gathered momentum until it drowned out the thunder itself and ended with a shuddering crash—followed by a sudden stillness. It took me a moment to realize we had heard part of a canyon wall collapsing.

The next morning I came out of my tent to survey the scene. The canyons stood in cathedral stillness in the pre-dawn light. The creek was back down to a trickle. The freshly washed sky arched overhead, so deep blue that you imagined you could see the empty vault of space beyond. The sun was just rising, and a band of scarlet light suddenly fired up the rim opposite our camp. Birds flitted about the mesquite trees. Breathing deeply of the fragrant air, I strolled up the canyon, looking for the landslide from the night before. But the canyon was already so thoroughly wrecked, every wall scarred with multiple landslides, that I could not identify the new one from the others.

According to an old Apache legend, when the Great Creator had finished making the earth, he had a heap of rejected rock and sand left over. Not knowing what else to do with it, he hurled it into the farthest corner of creation. It came to rest in a senseless jumble—thus creating the Big Bend. Standing there, I knew that the Great Creator wasn't finished with this landscape and never would be. It is a work in progress, a slow sculpture of infinite complexity. In the end it is a place where we human beings will always be a humble and fleeting presence.

Opposite page: Ribbons of time.

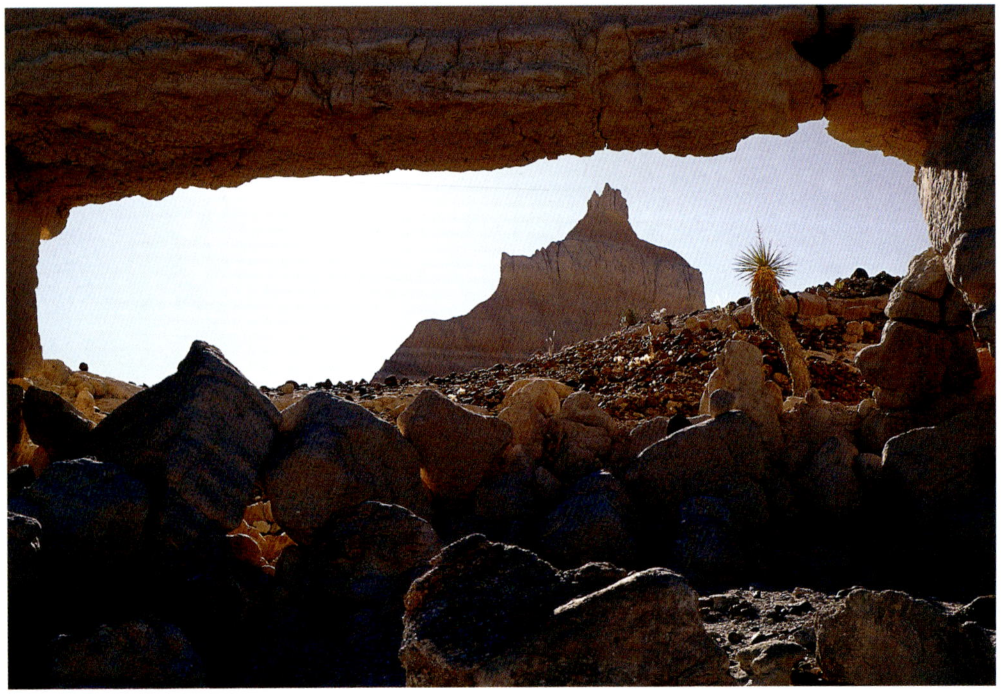

This page: Wedge viewed through cathedral arch.
Opposite page: Weathering in the lower mezzanine.

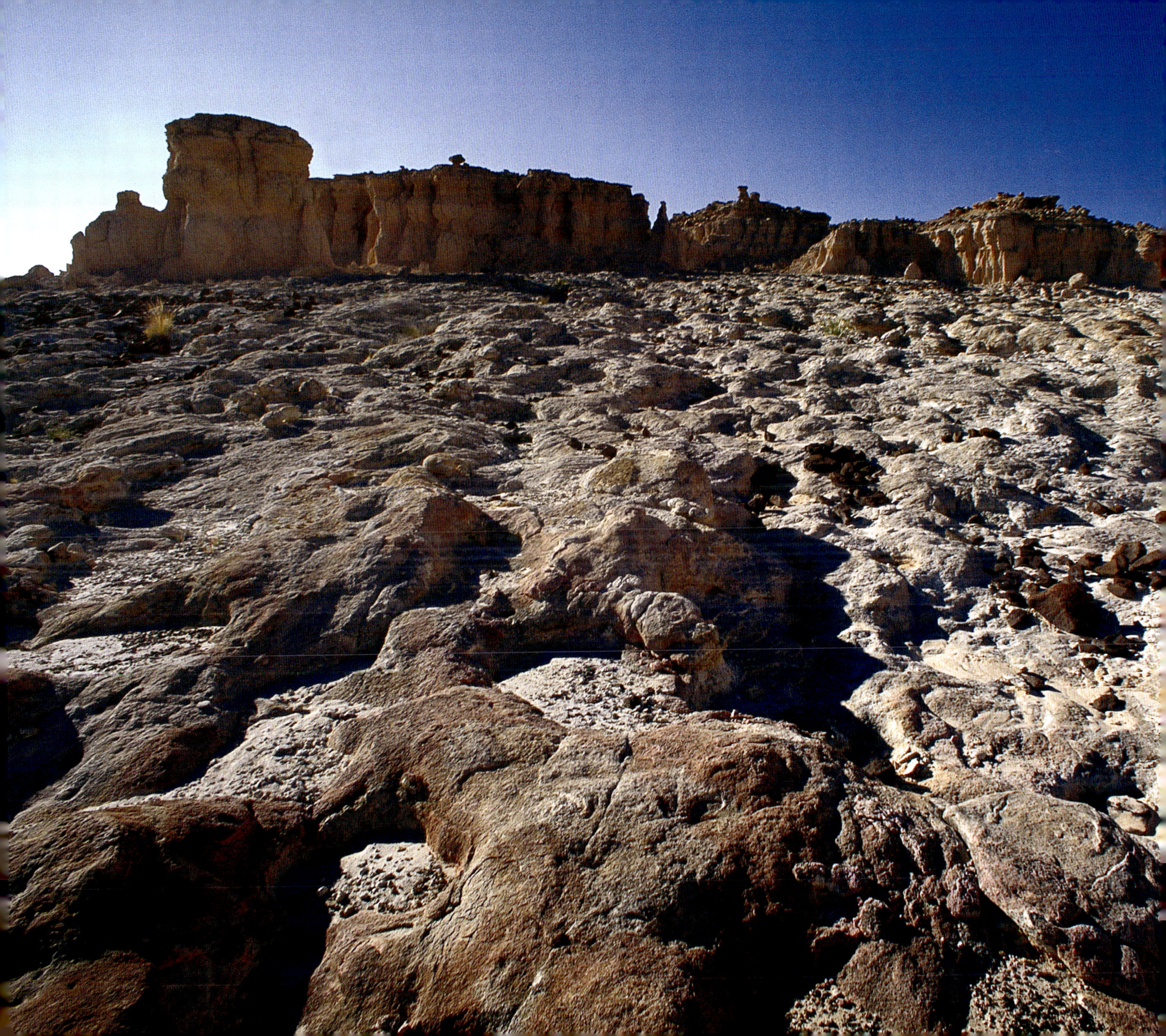

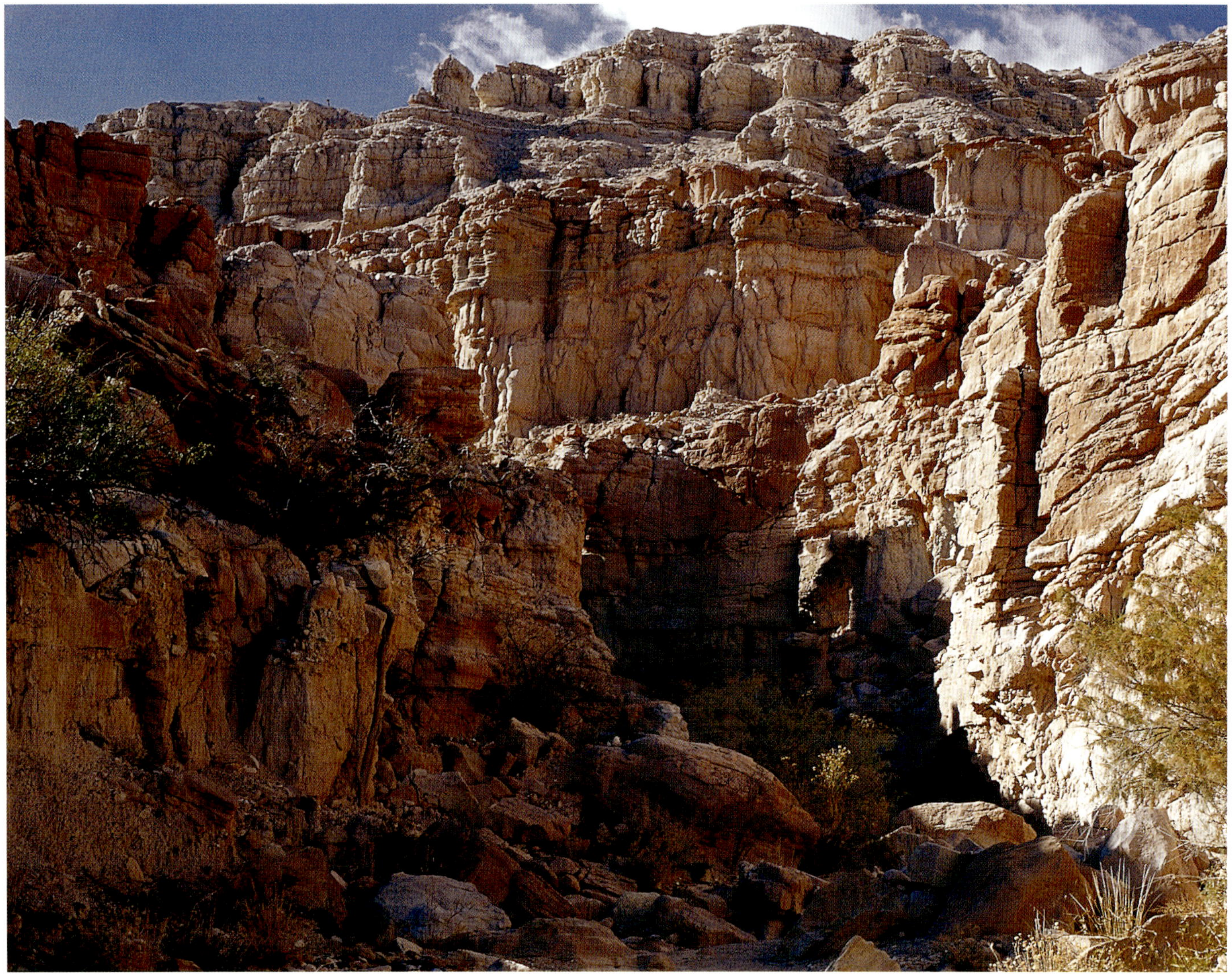

This page: White canyon entrance.
Opposite page: Rose-tinted sediments below hoodoo, base of Dalquest peak.

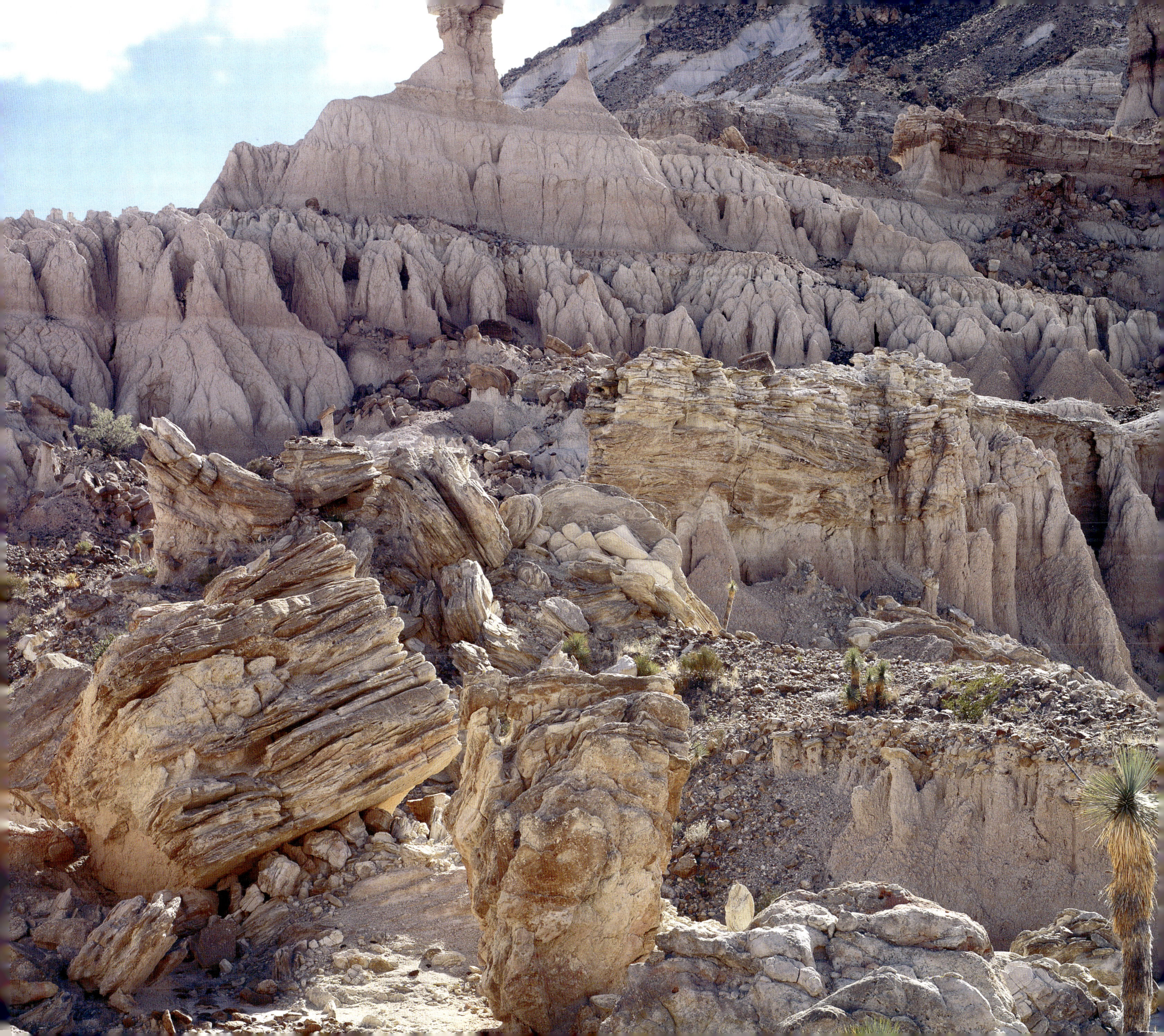

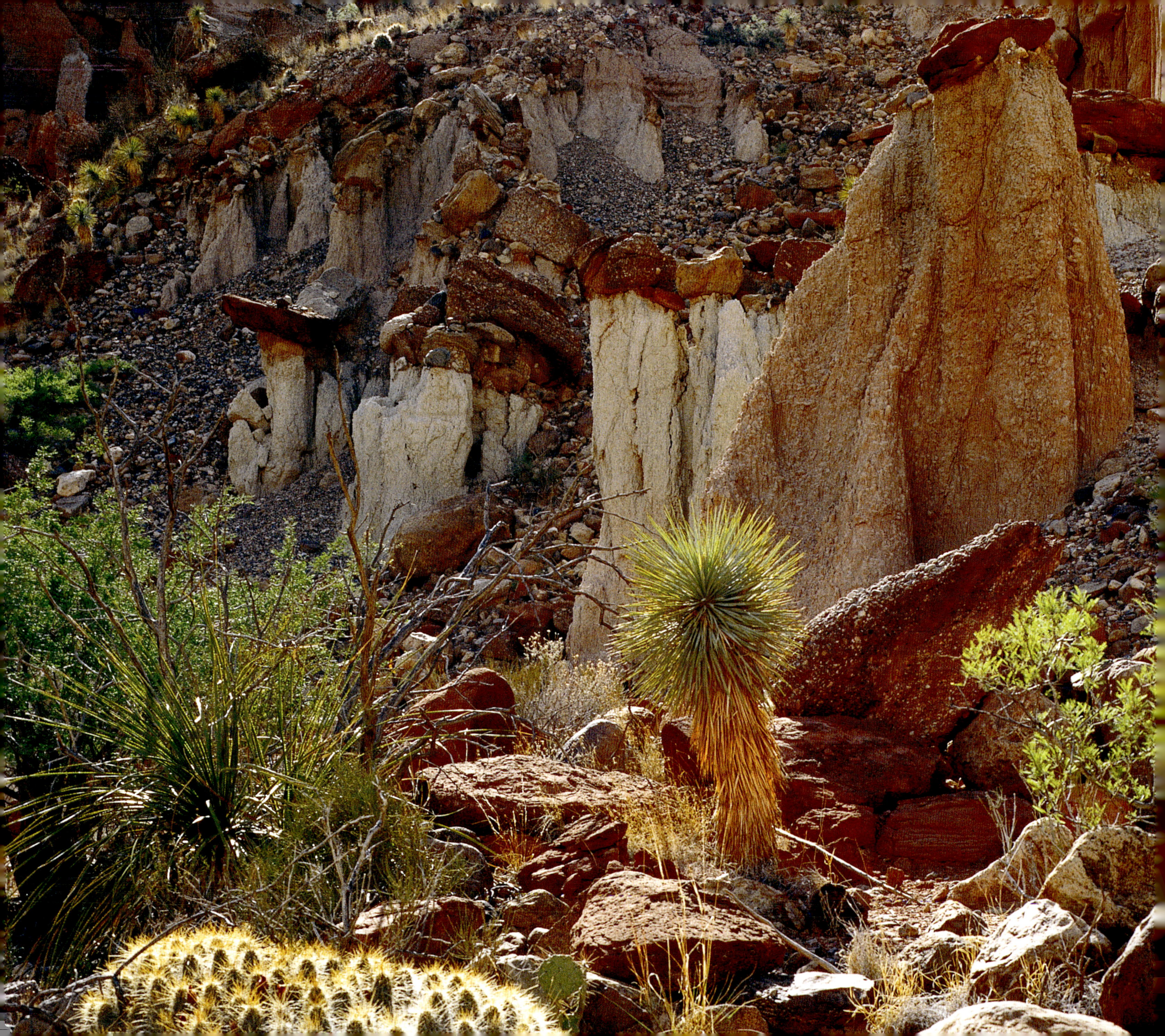

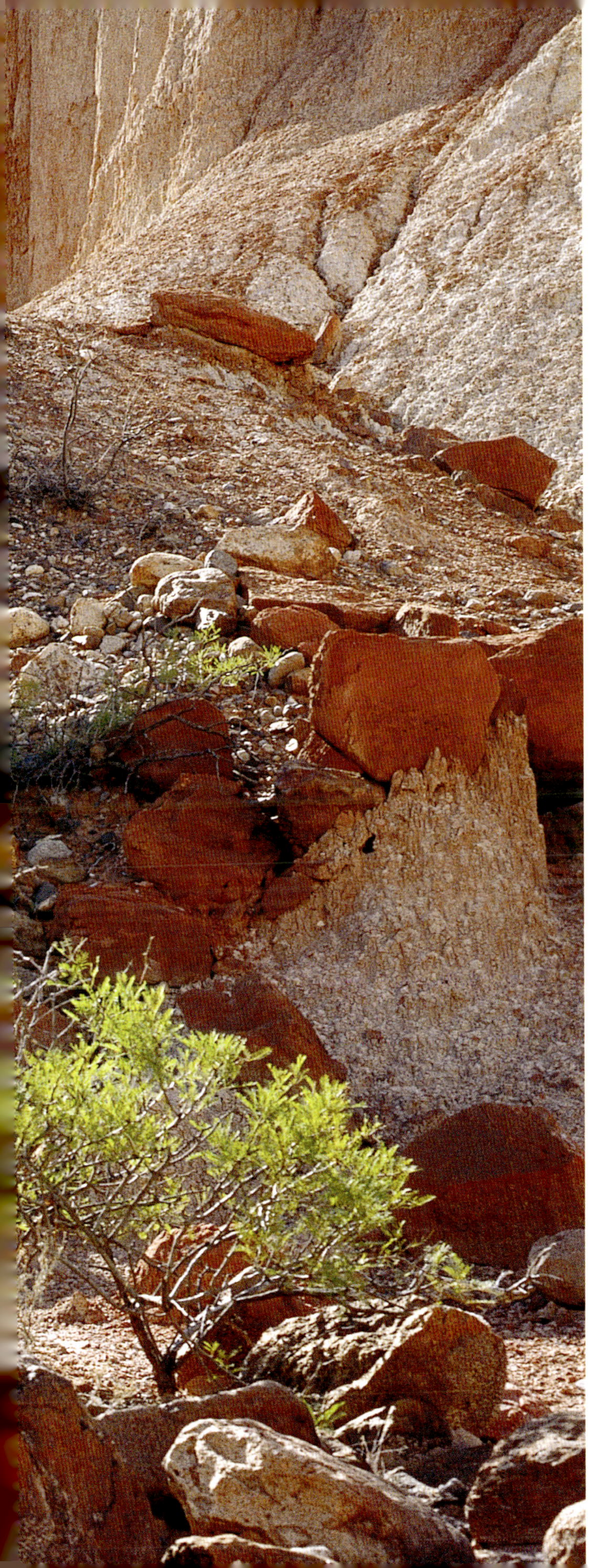

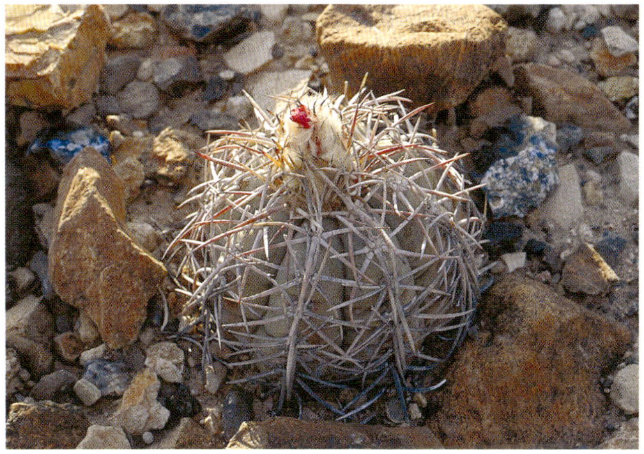

Opposite page: Slot canyon off red canyon.
This page: Devil's head cactus

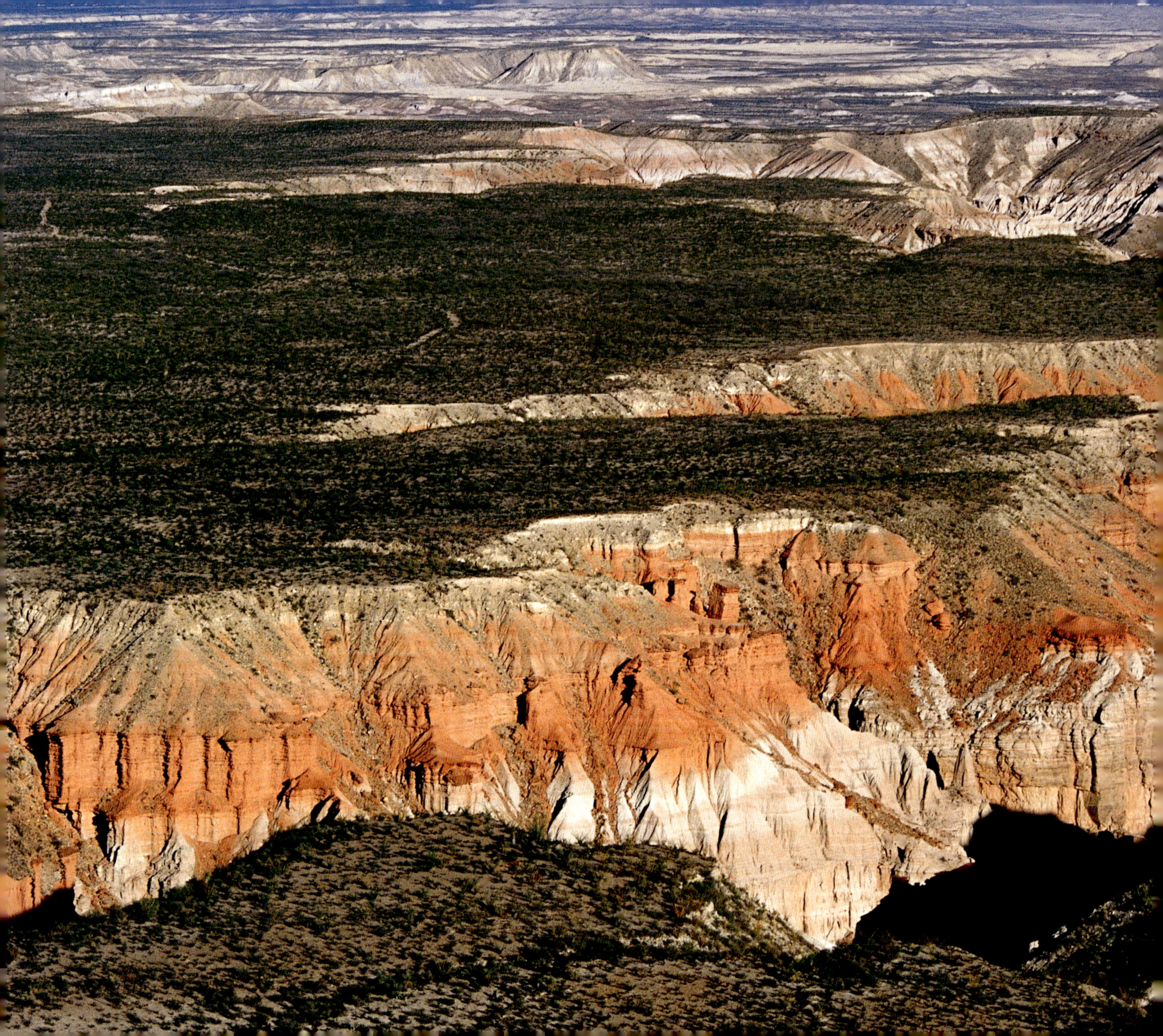

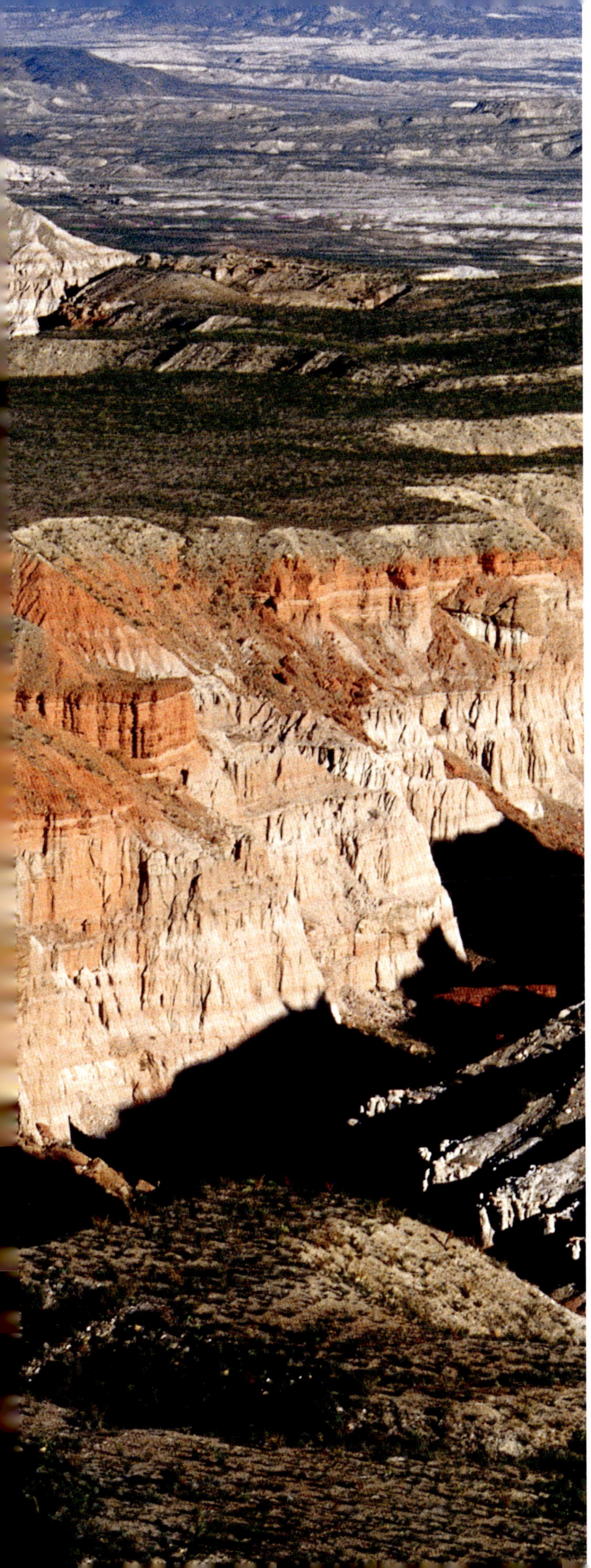

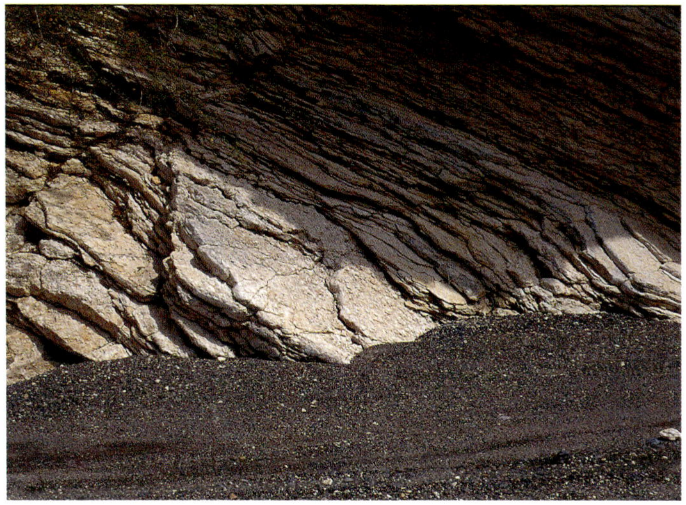

Opposite page: Bandera Mesa beyond red canyon.

This page: Igneous rock covered by basalt gravel wash.

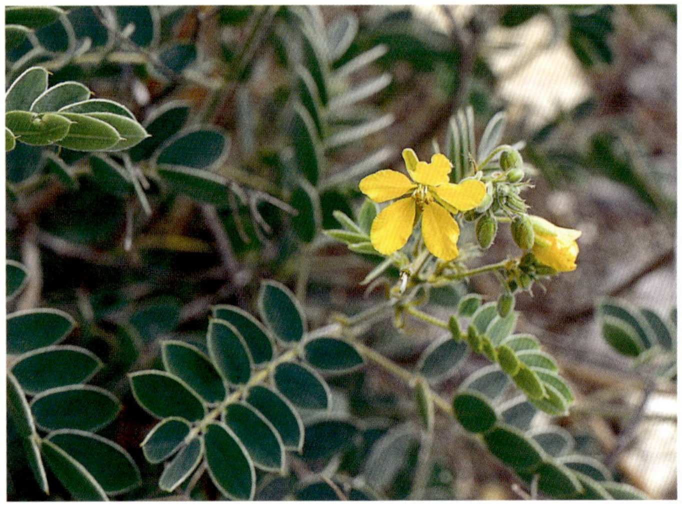

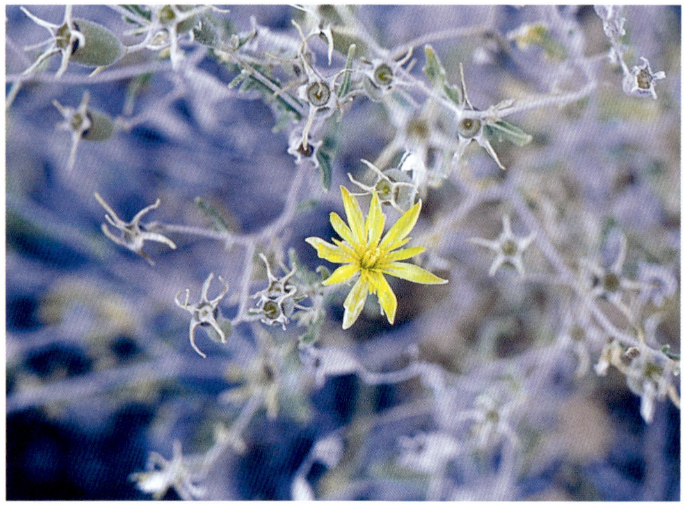

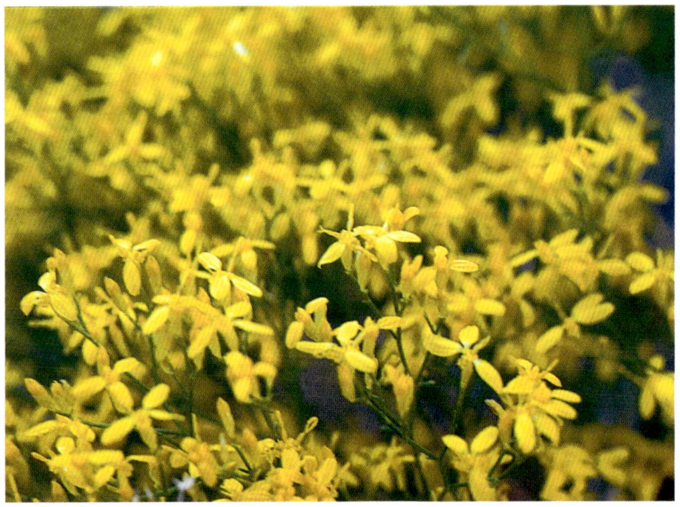

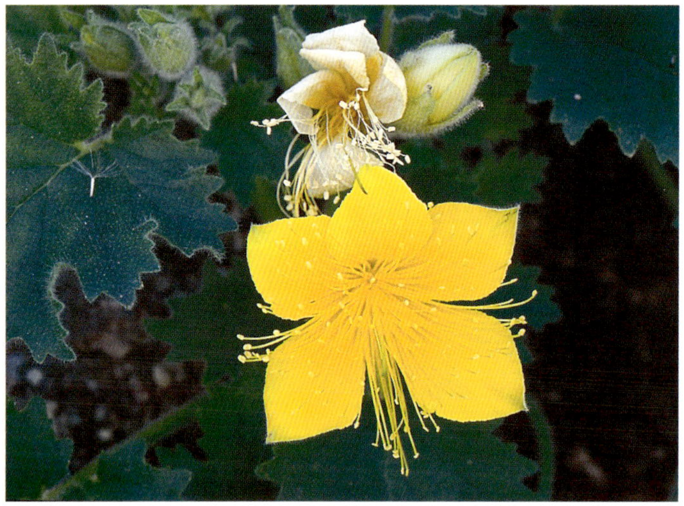

Opposite page: Reflections in red canyon.
This page, above left: Yellow senna.
Above right: Prairie stickleaf.
Below left: Snakeweed.
Below right: Yellow rocknettle.

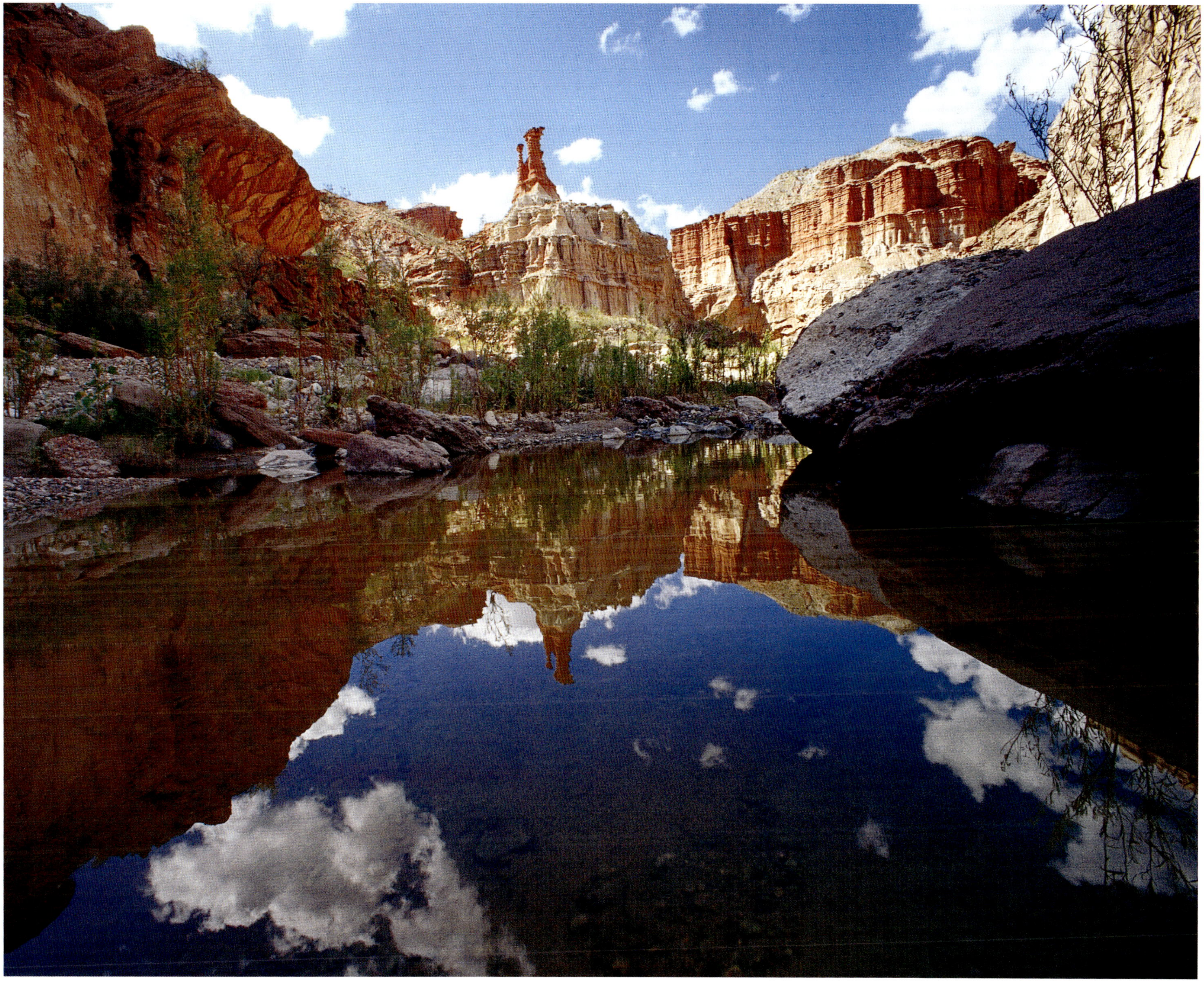

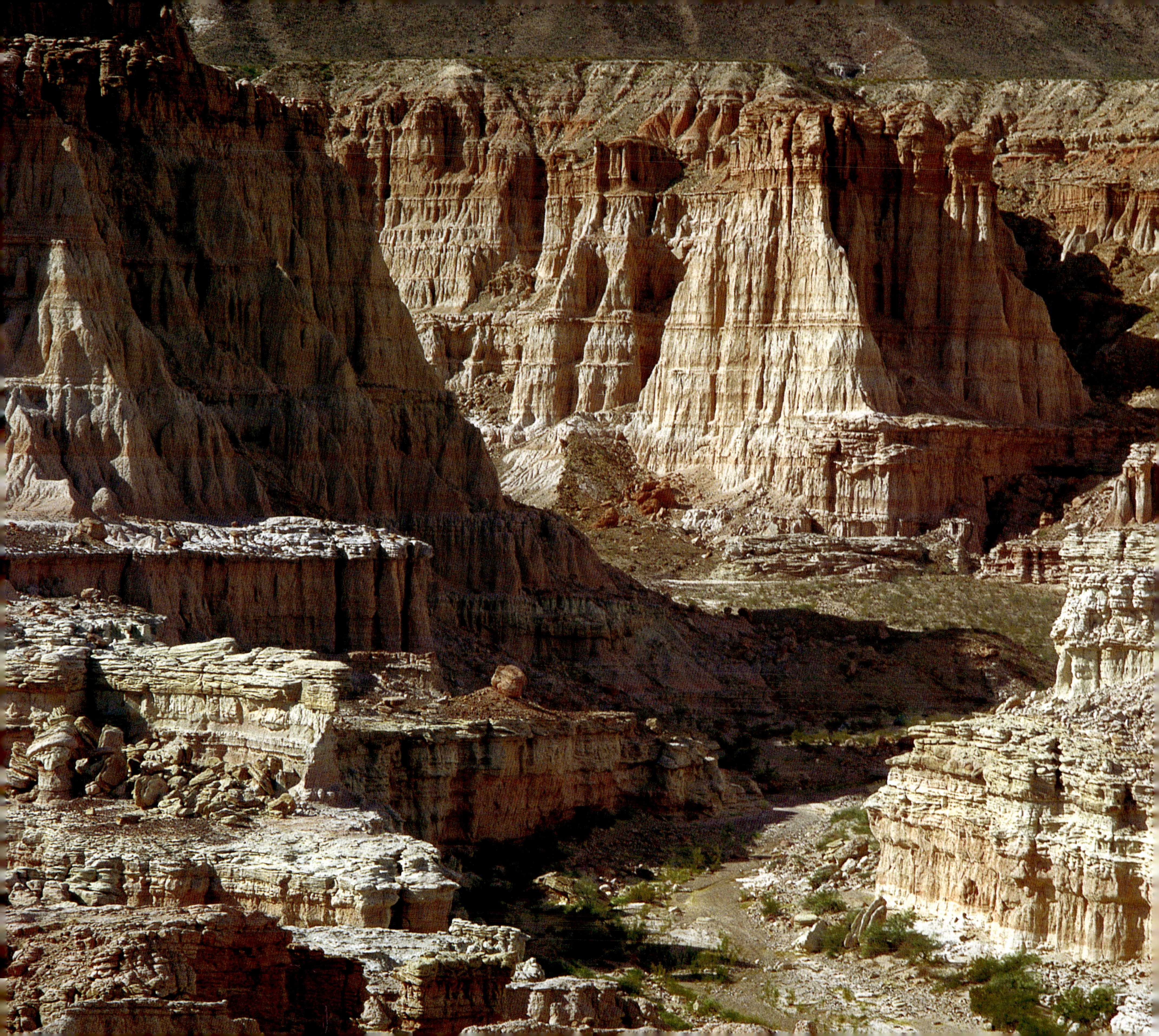

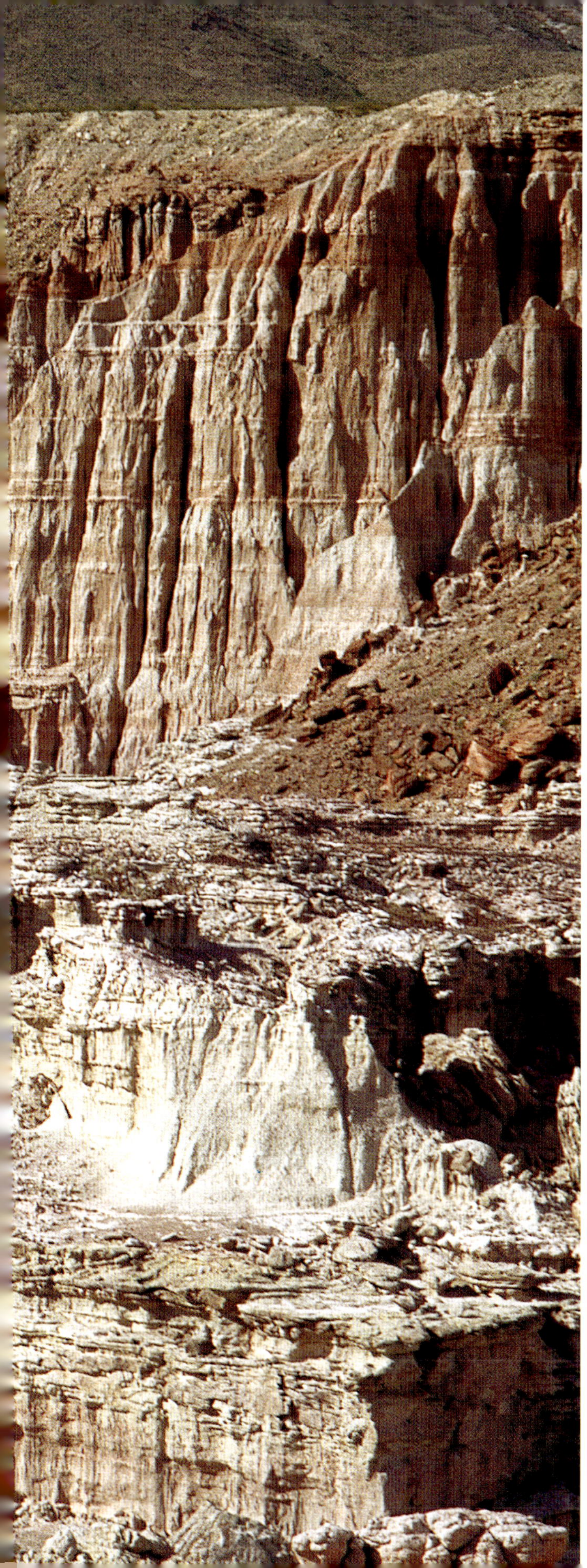

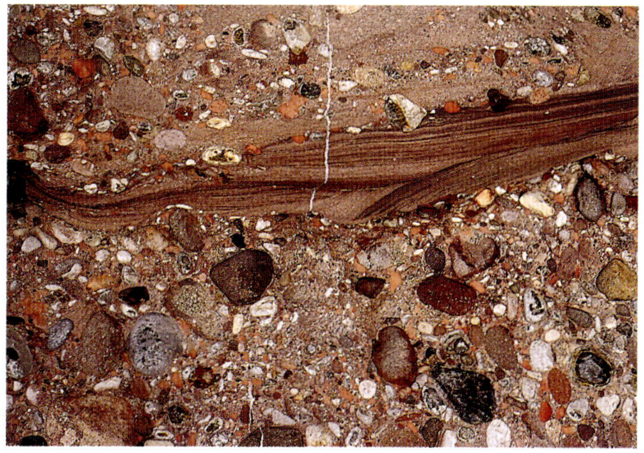

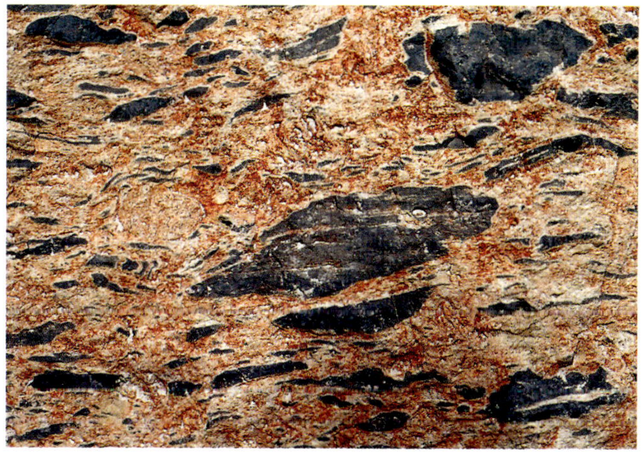

Opposite page: Walls of red canyon.

This page, above: Conglomerate cut by sandstone lens.

Below: Volcanic breccia.

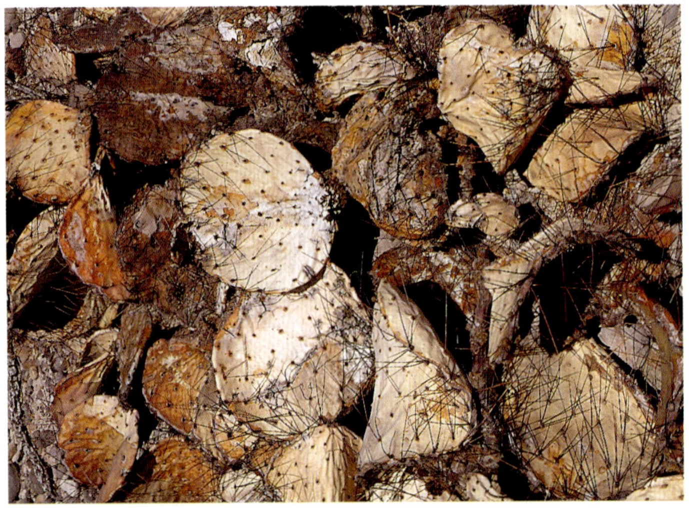

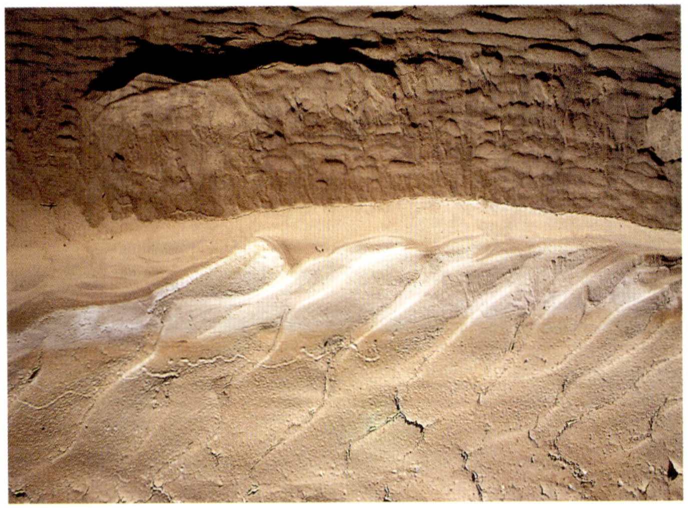

This page, above: Decaying prickly pear pads.
Below: Ripple marks on canyon floor.
Opposite page: Red canyon, looking west.

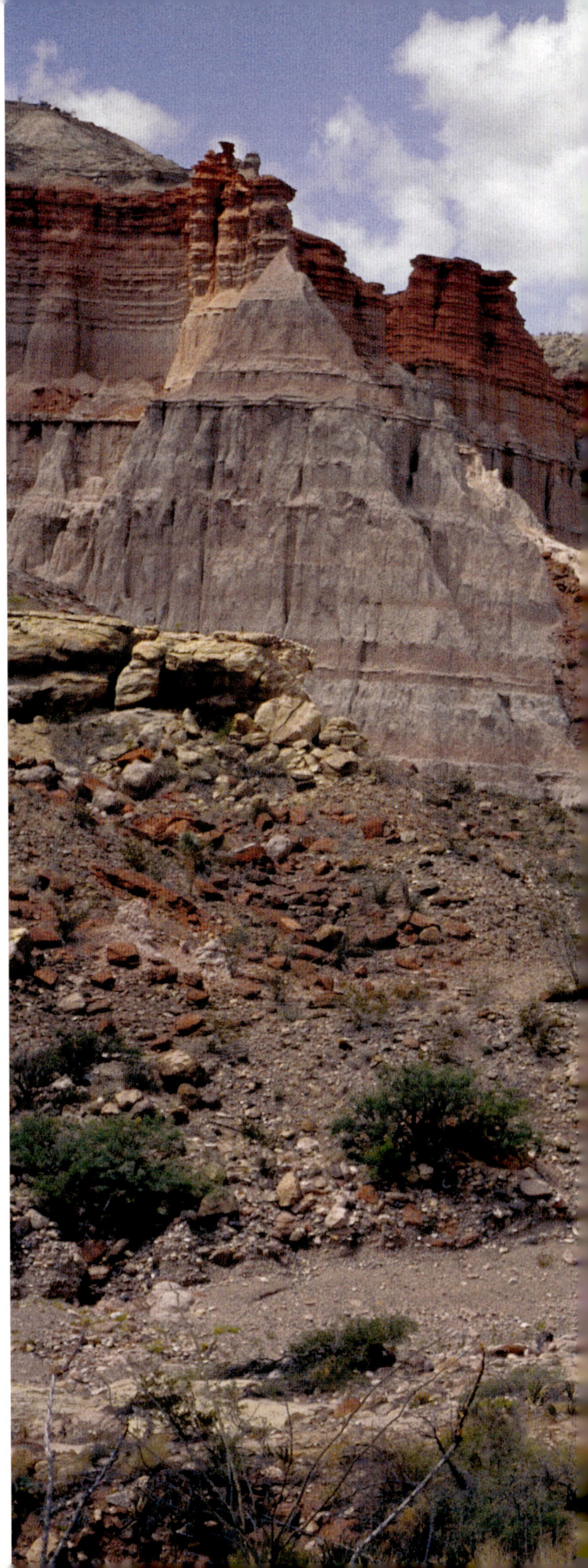

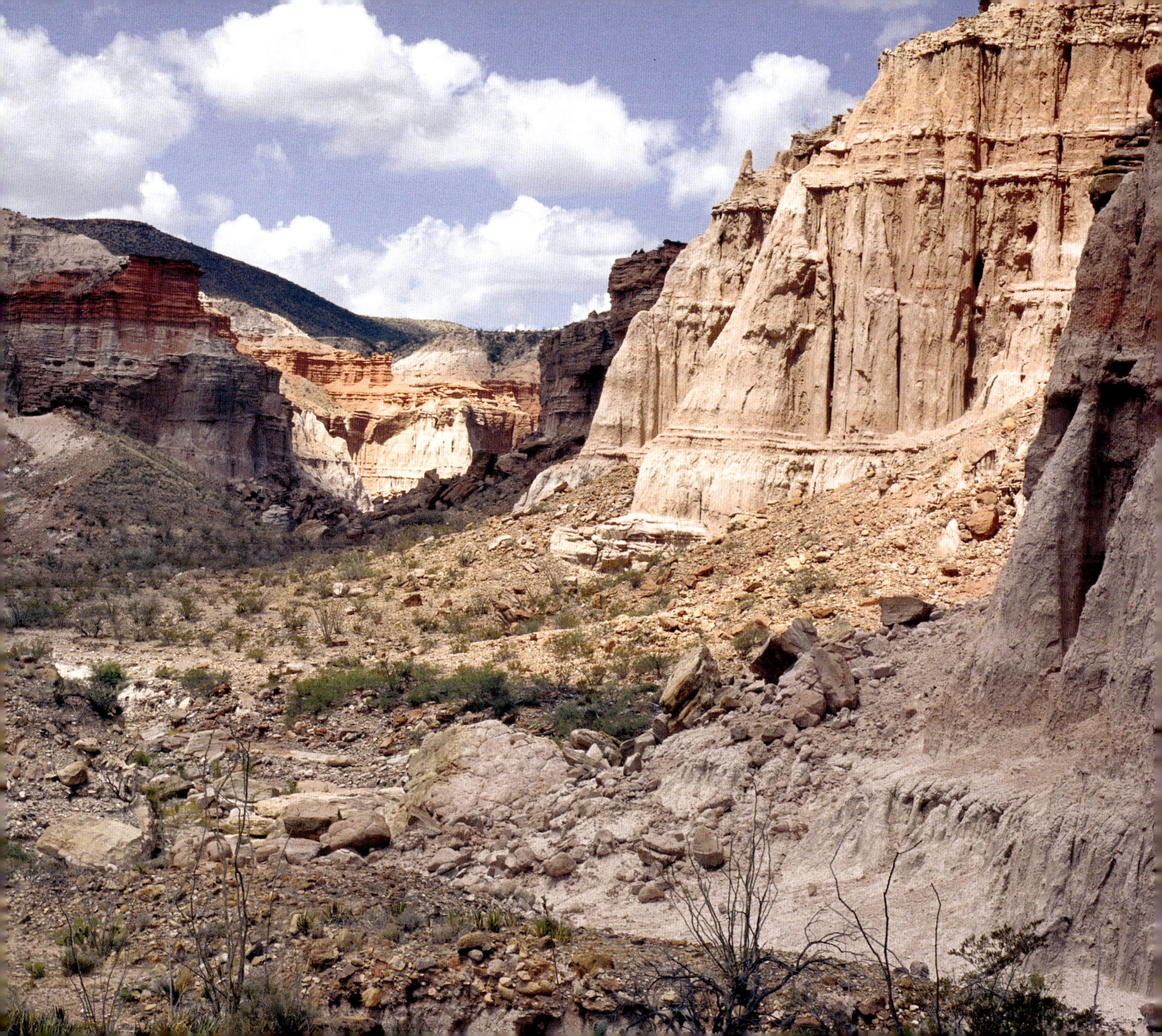

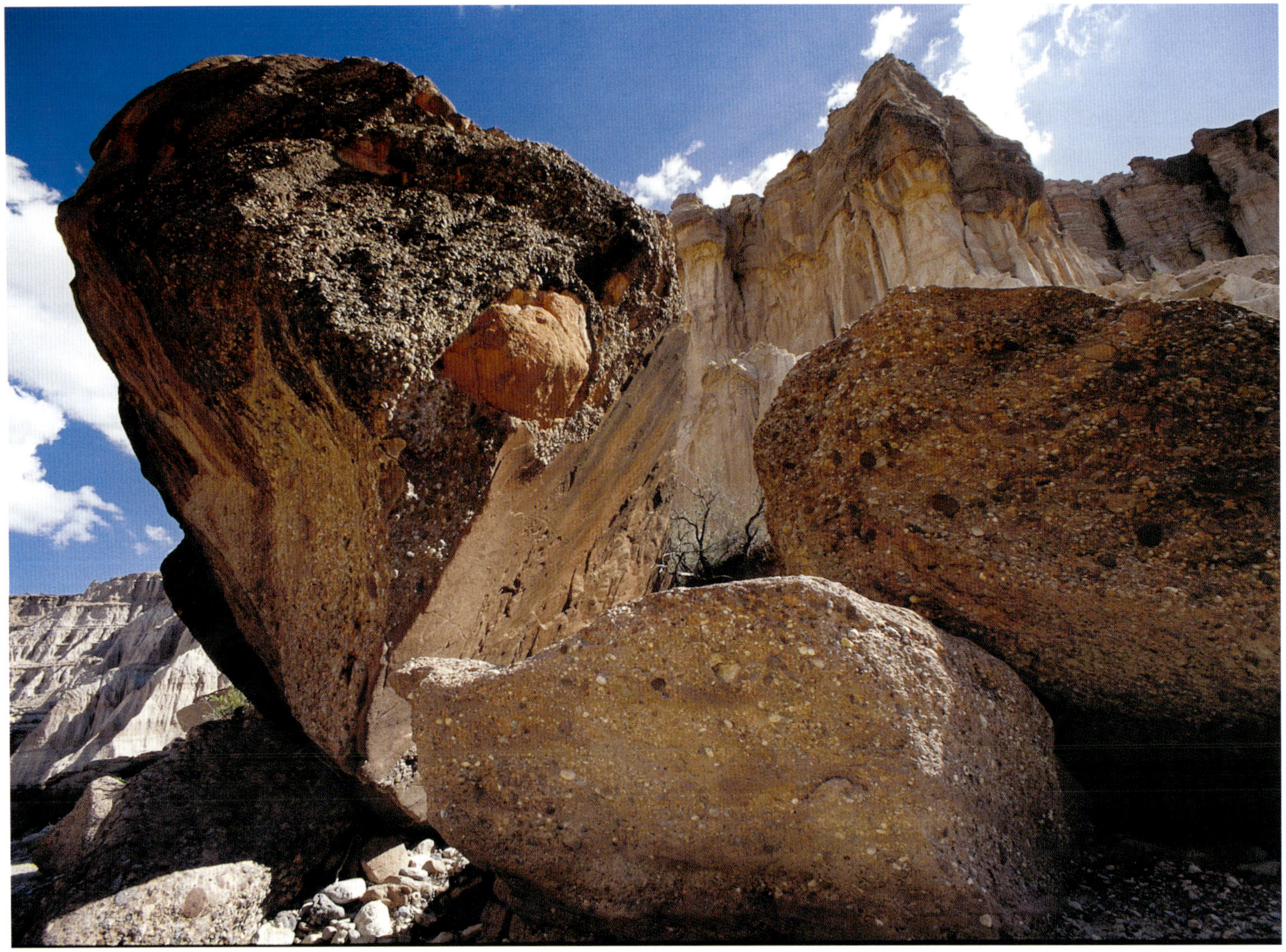

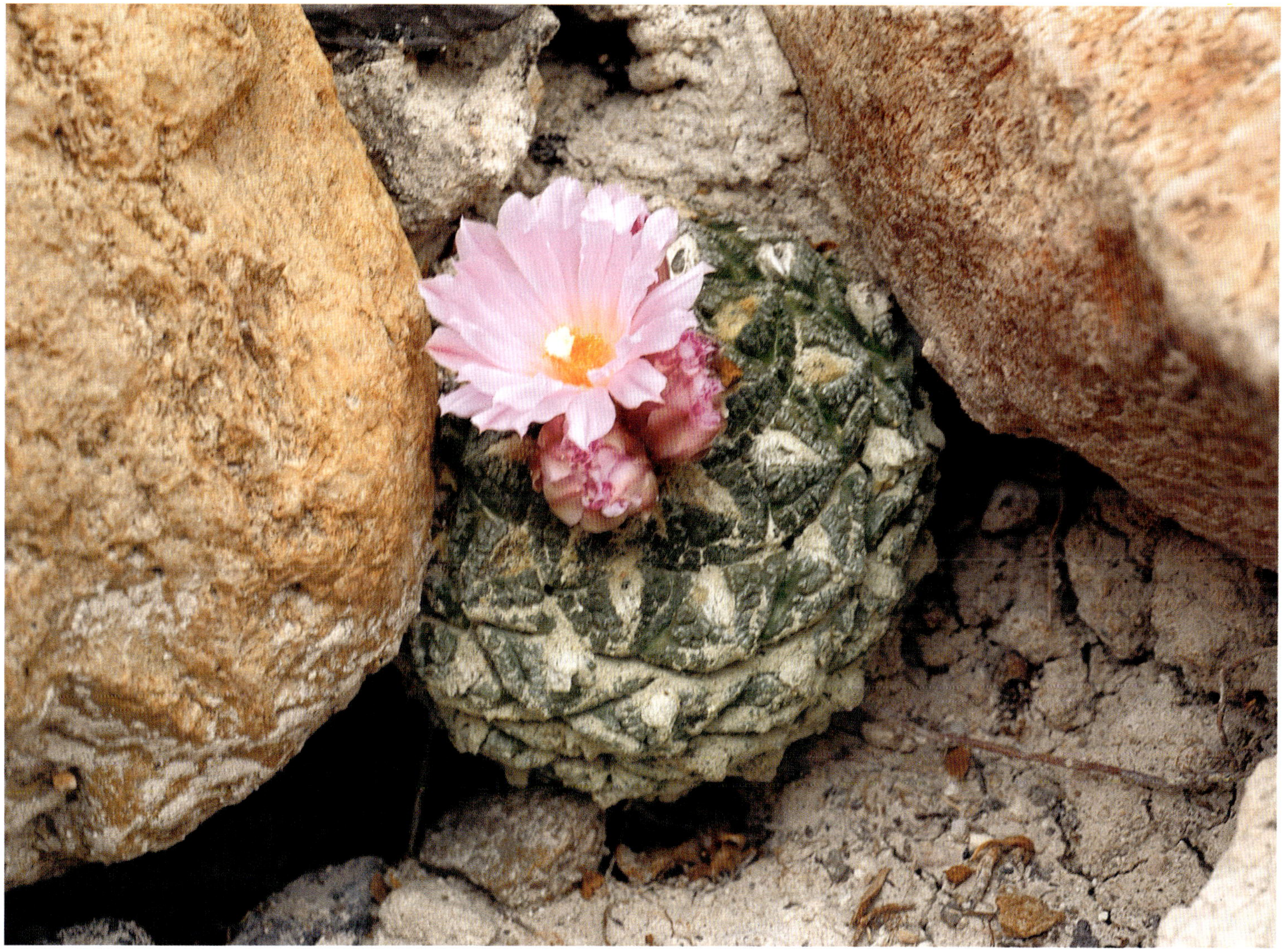

Opposite page: Conglomerate boulder fall.

*This page: Living rock cactus (*Ariocarpus fissuratus*).*

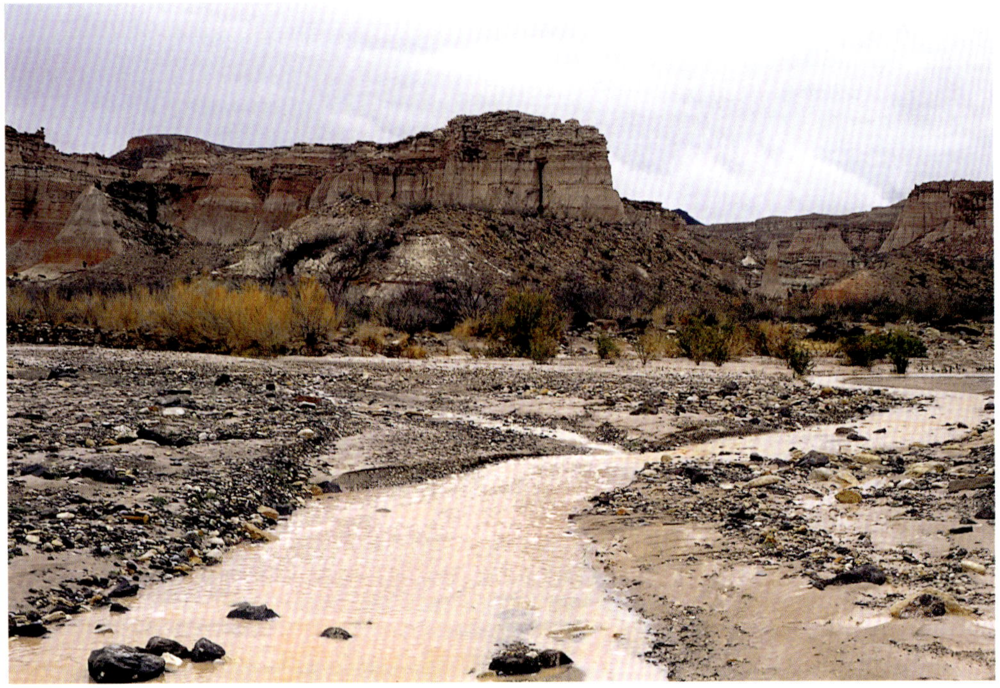

This page: North and south fork confluence,
Alamo de Cesario Creek.
Opposite page: Sunrise at red canyon.

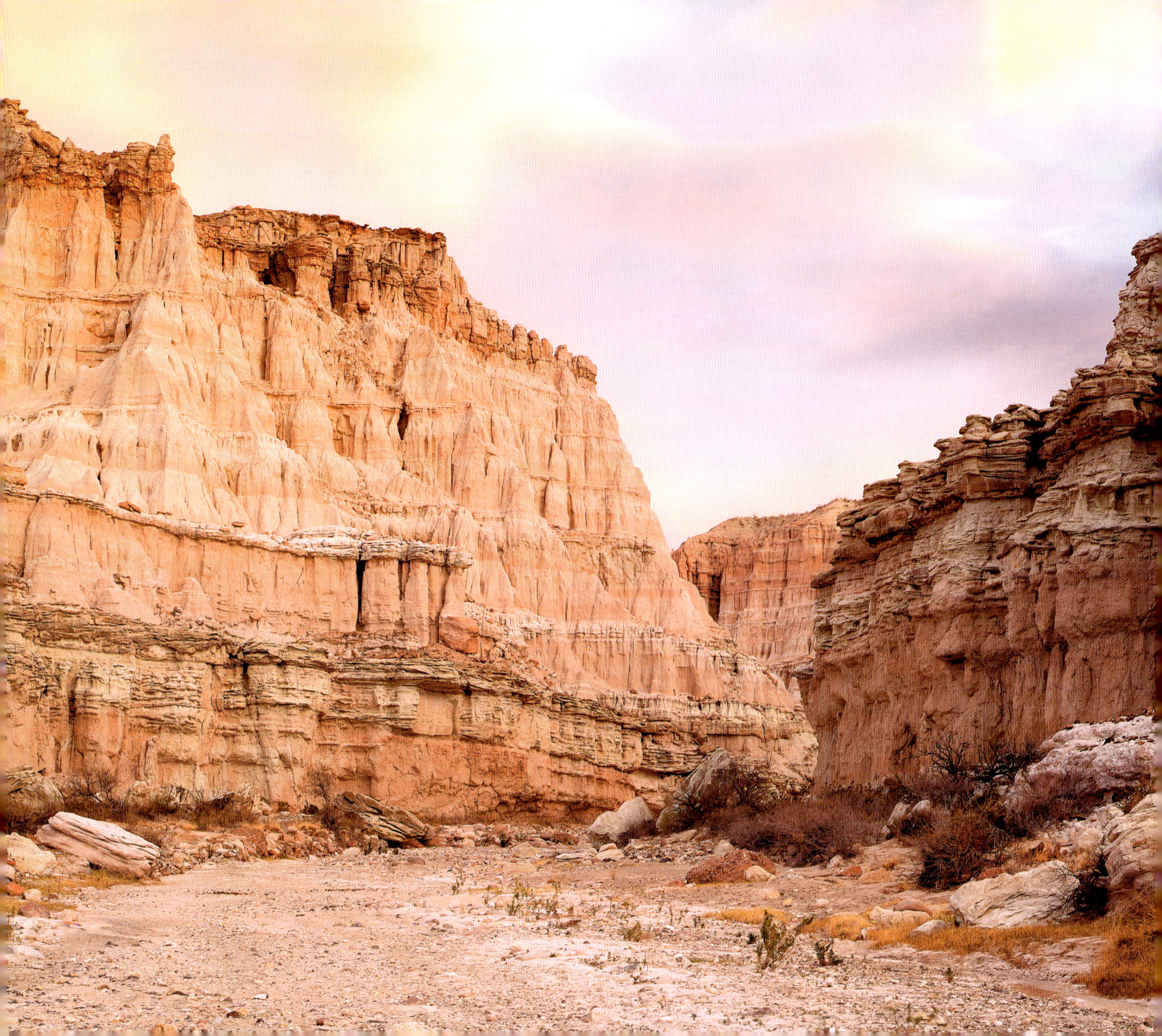

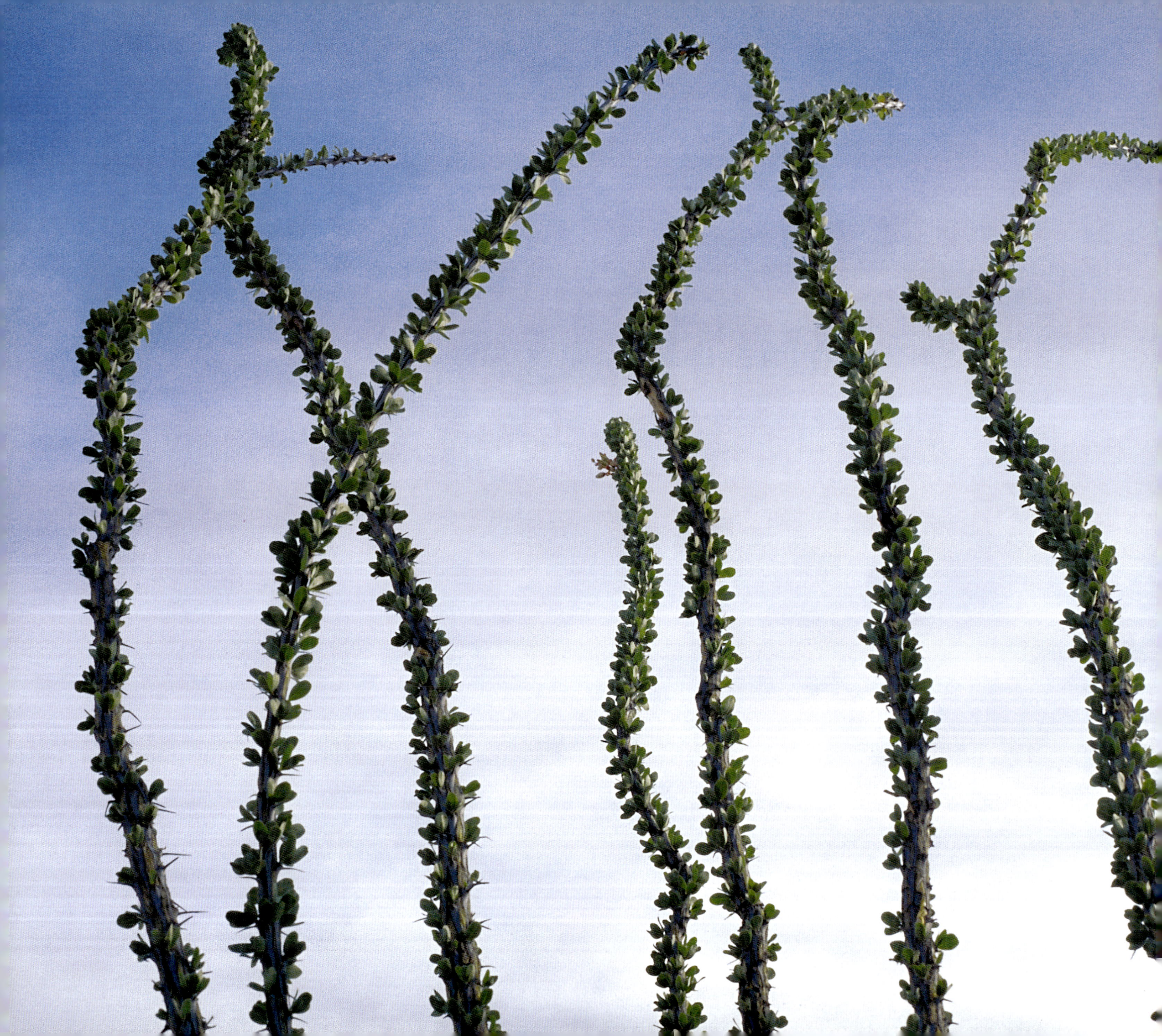

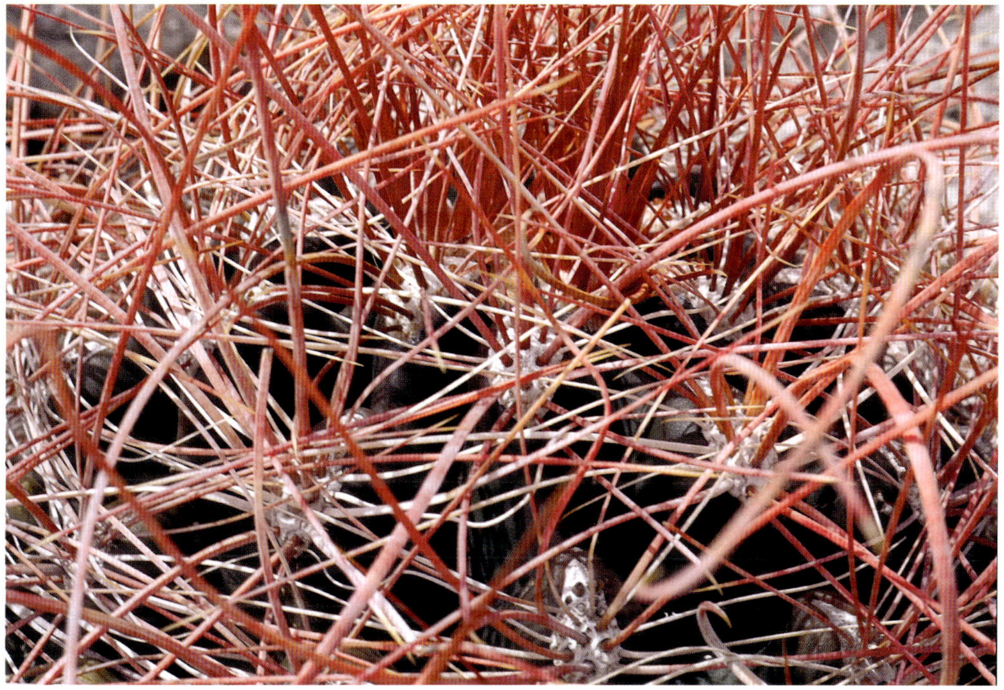

Opposite page: Ocotillo.

This page: Fishhook cactus.

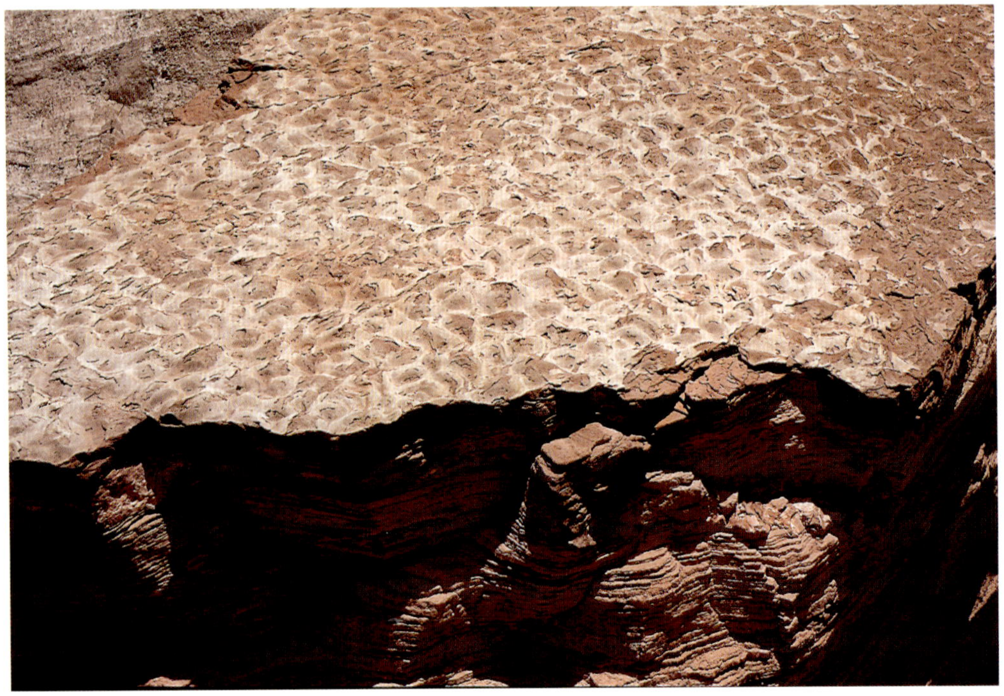

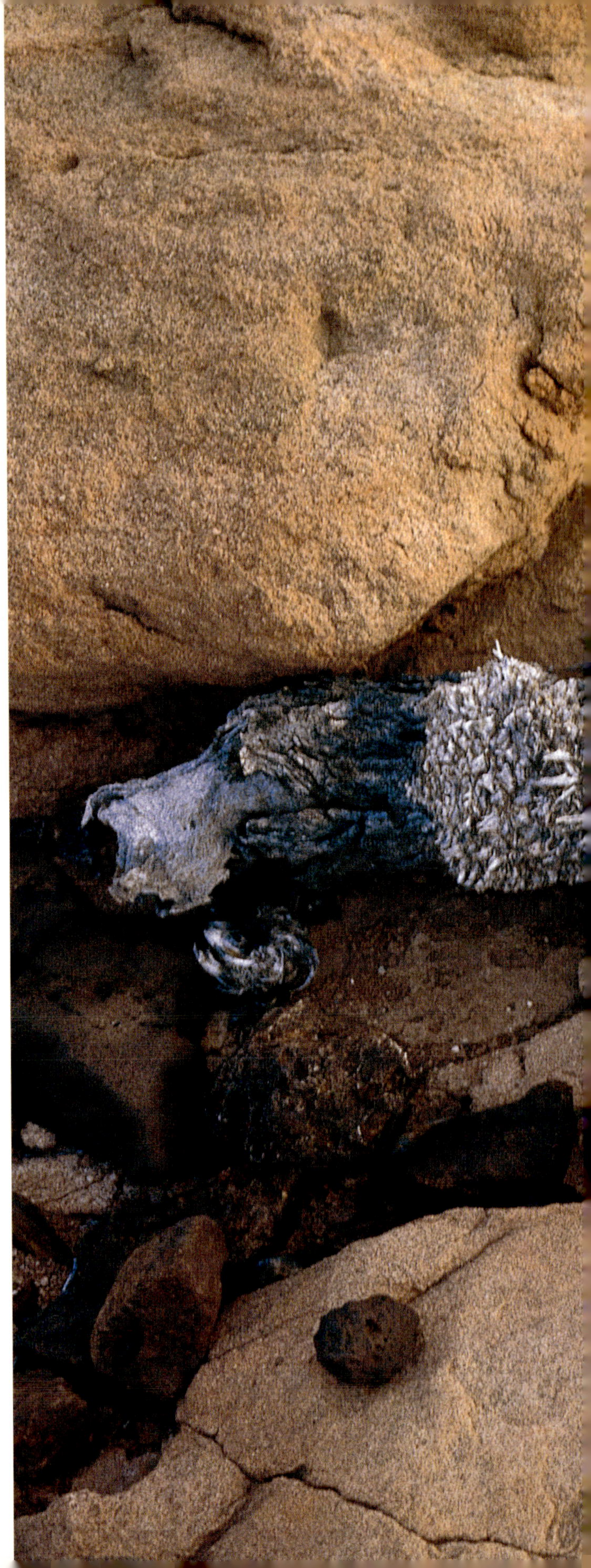

This page: Water-scalloped rock.
Opposite page: Yucca generations.

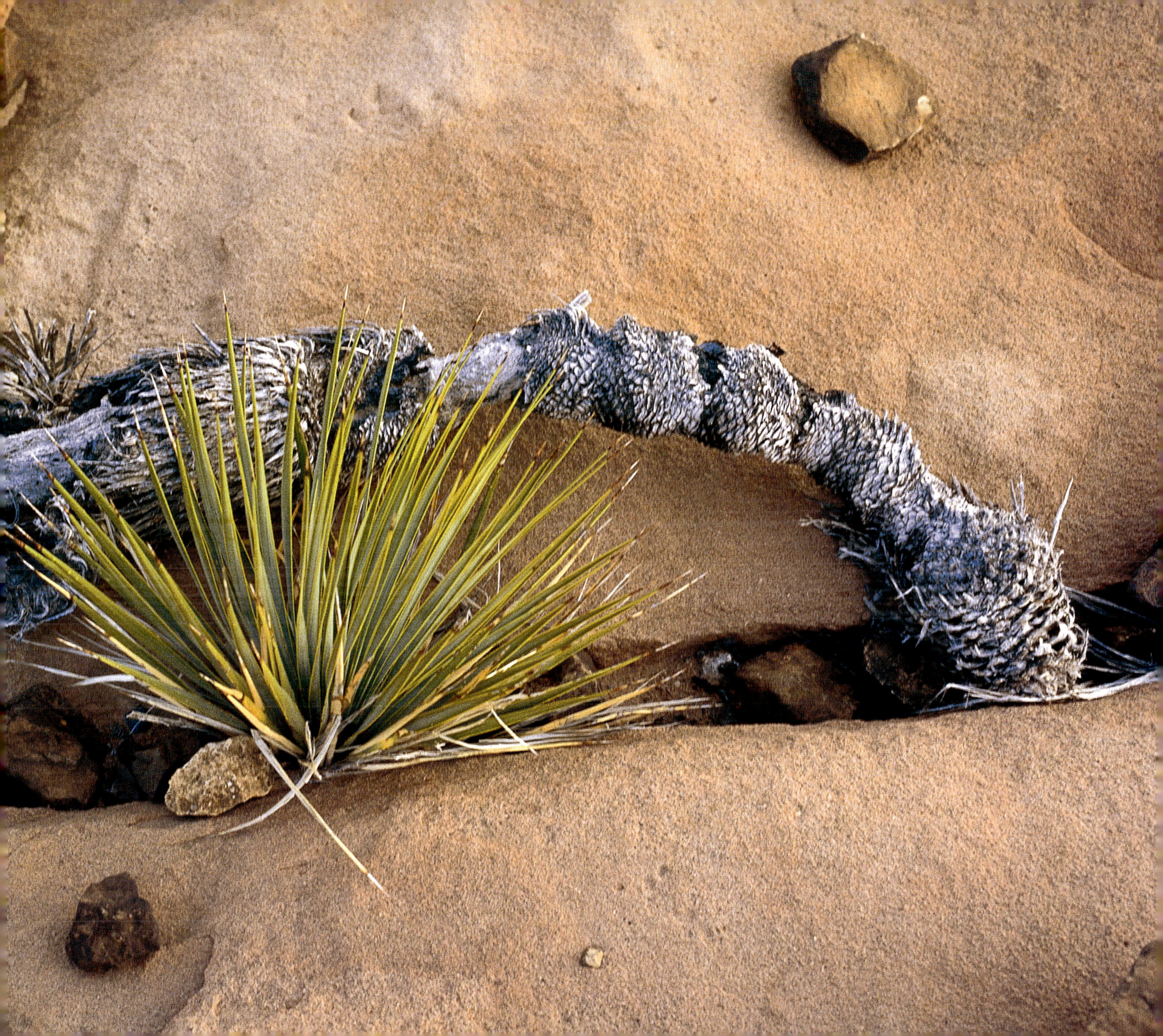

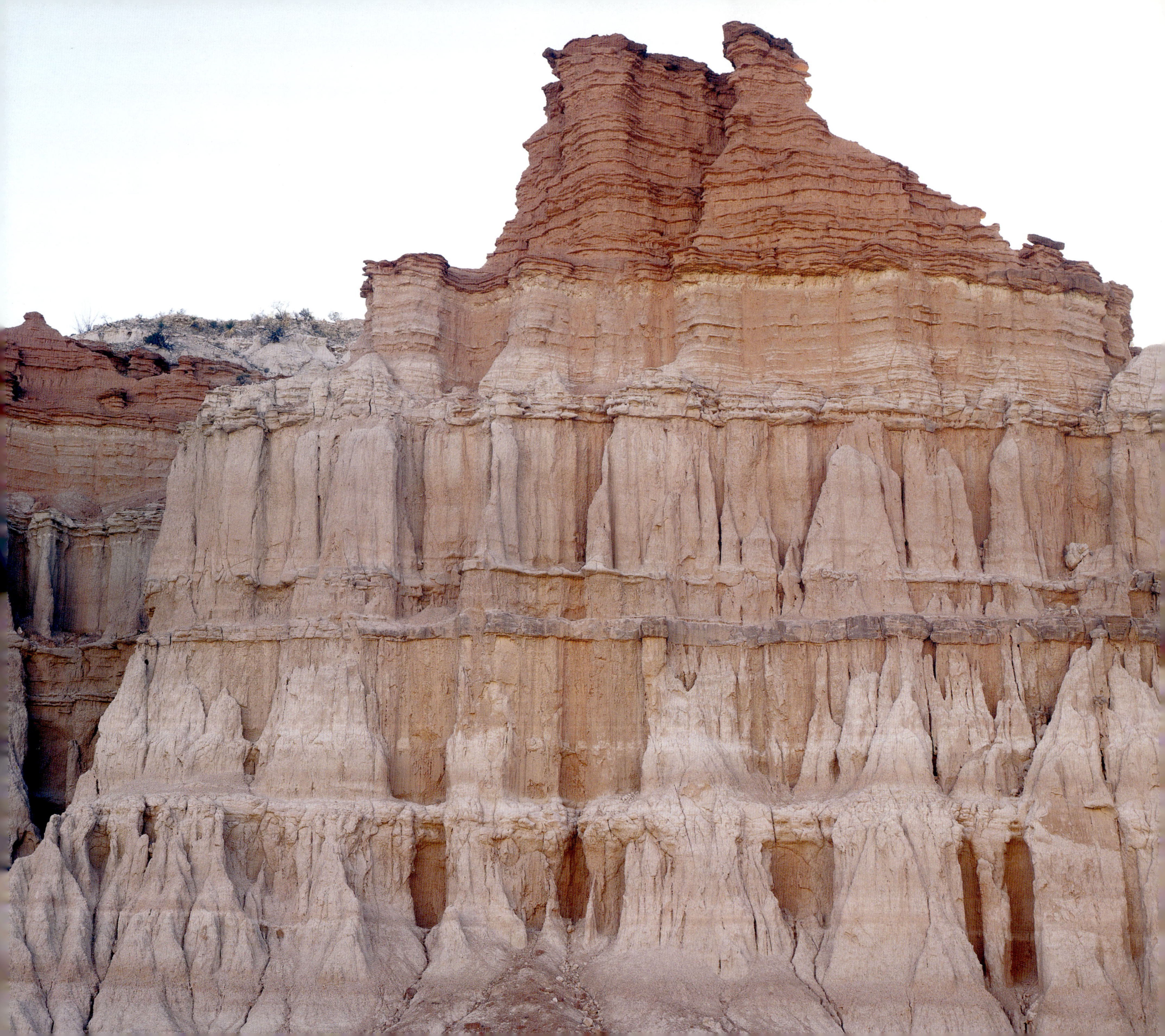

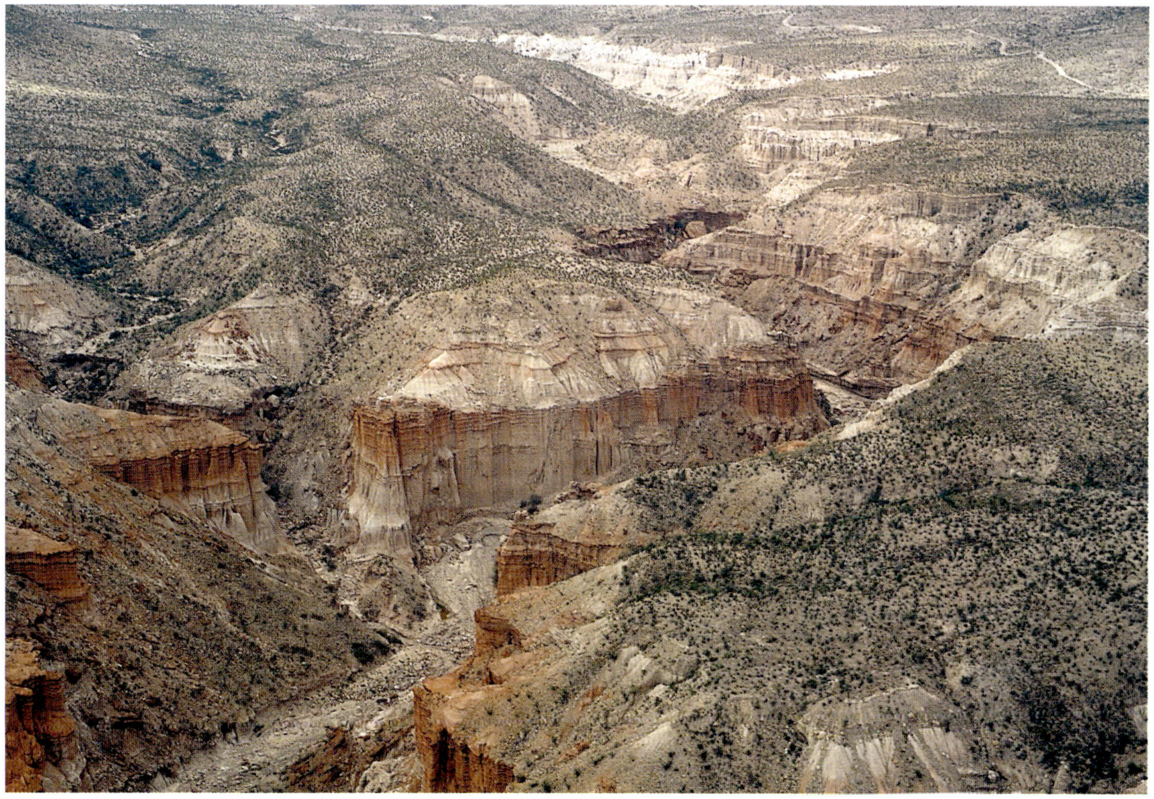

Opposite page: Duomo.
This page: Alamo de Cesario Canyon.

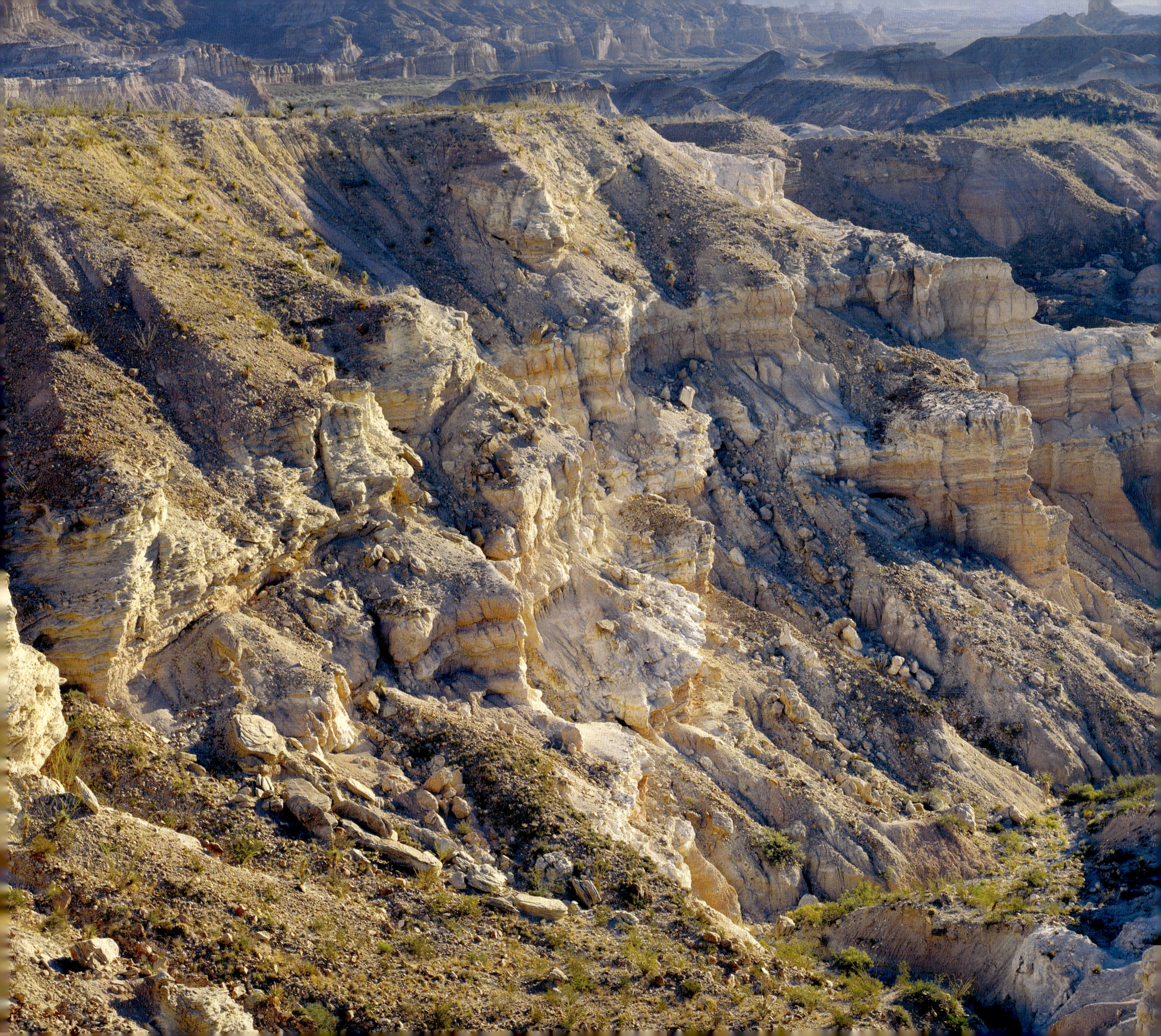

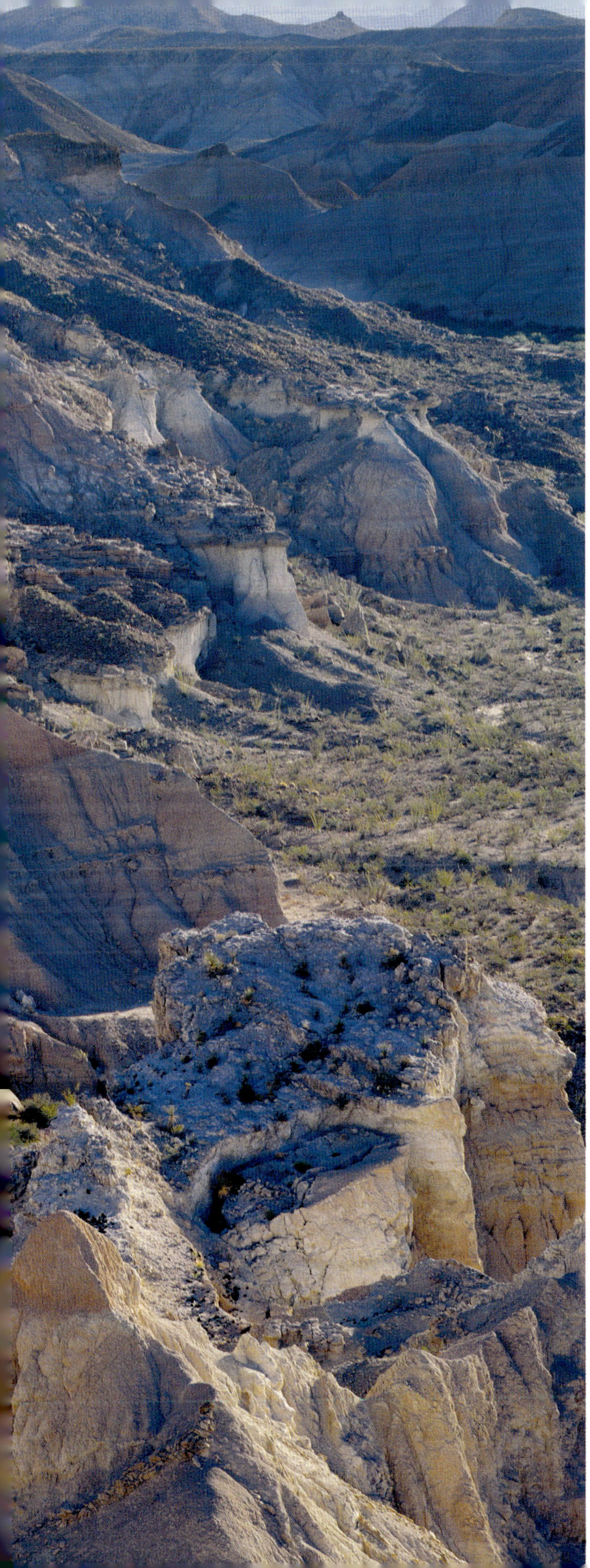

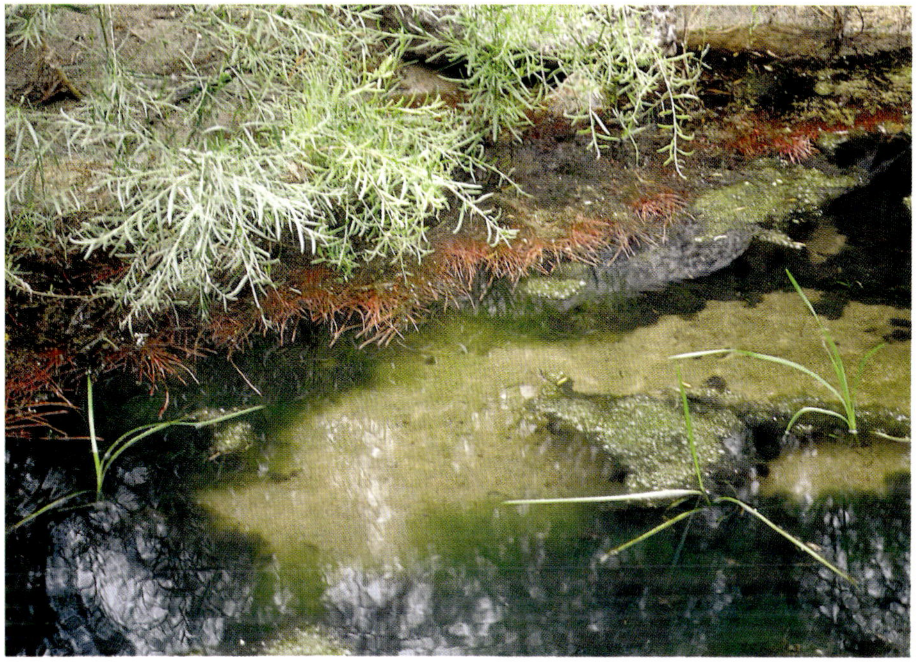

Opposite page: Results of weathering and erosion in the Devil's Graveyard.
This page: Desert pool ecosystem.

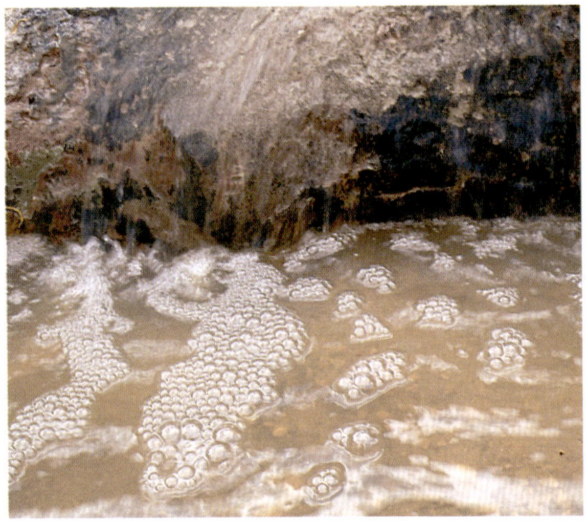
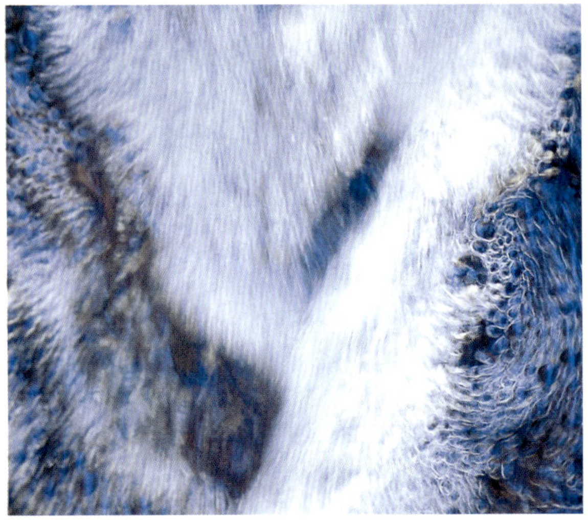
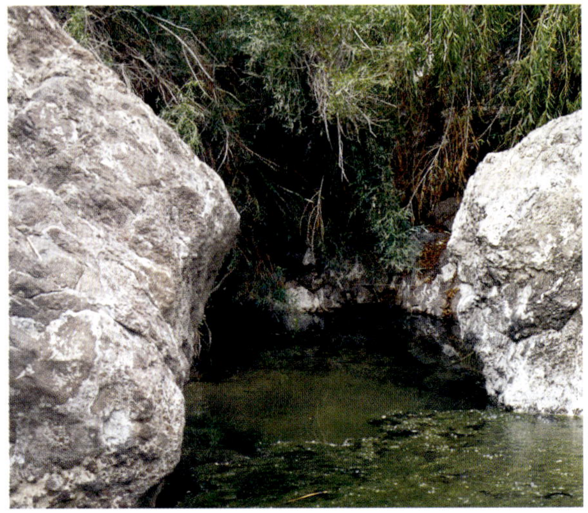

This page: Desert waters.
Opposite page: Water-carved sandstone.

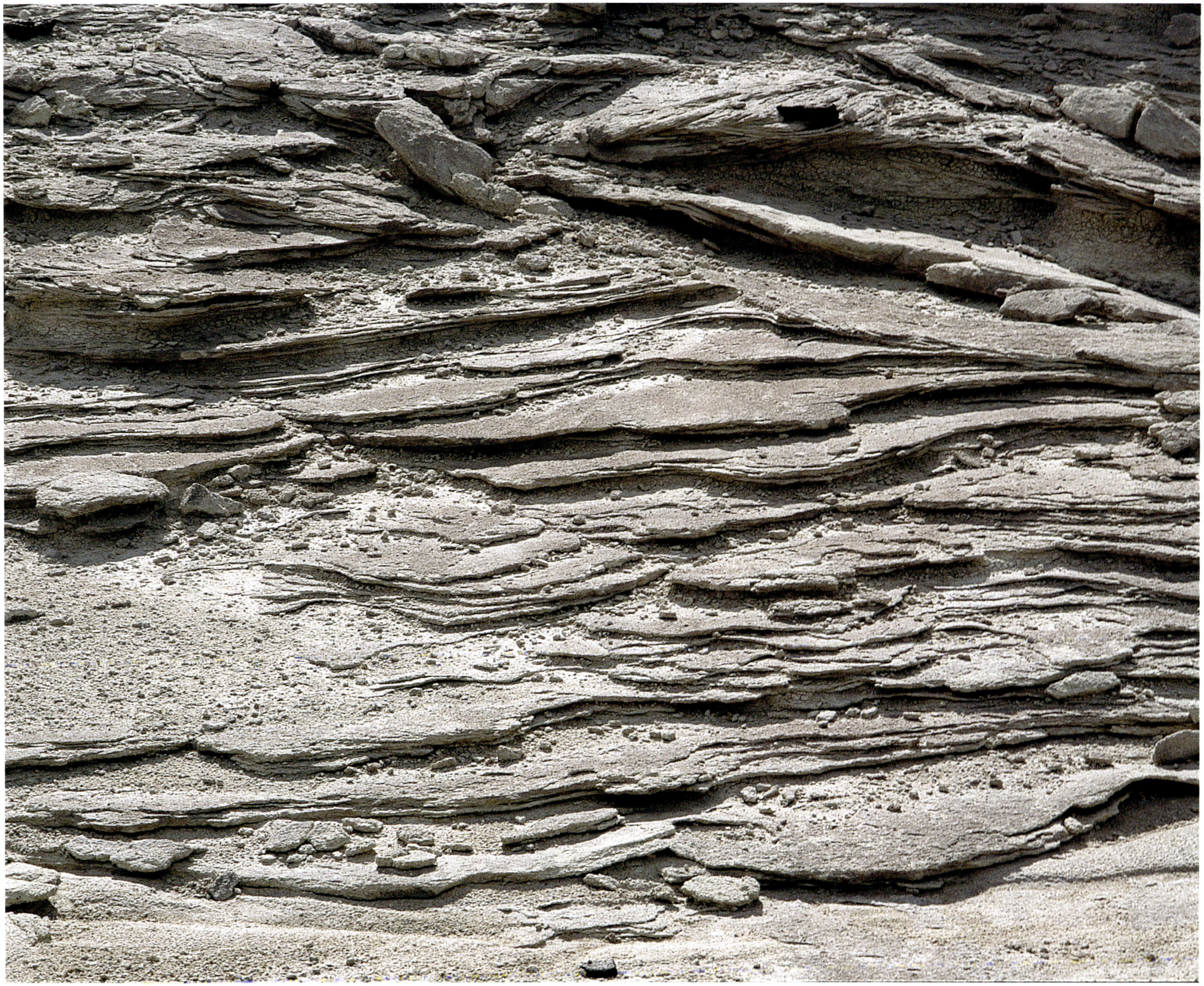

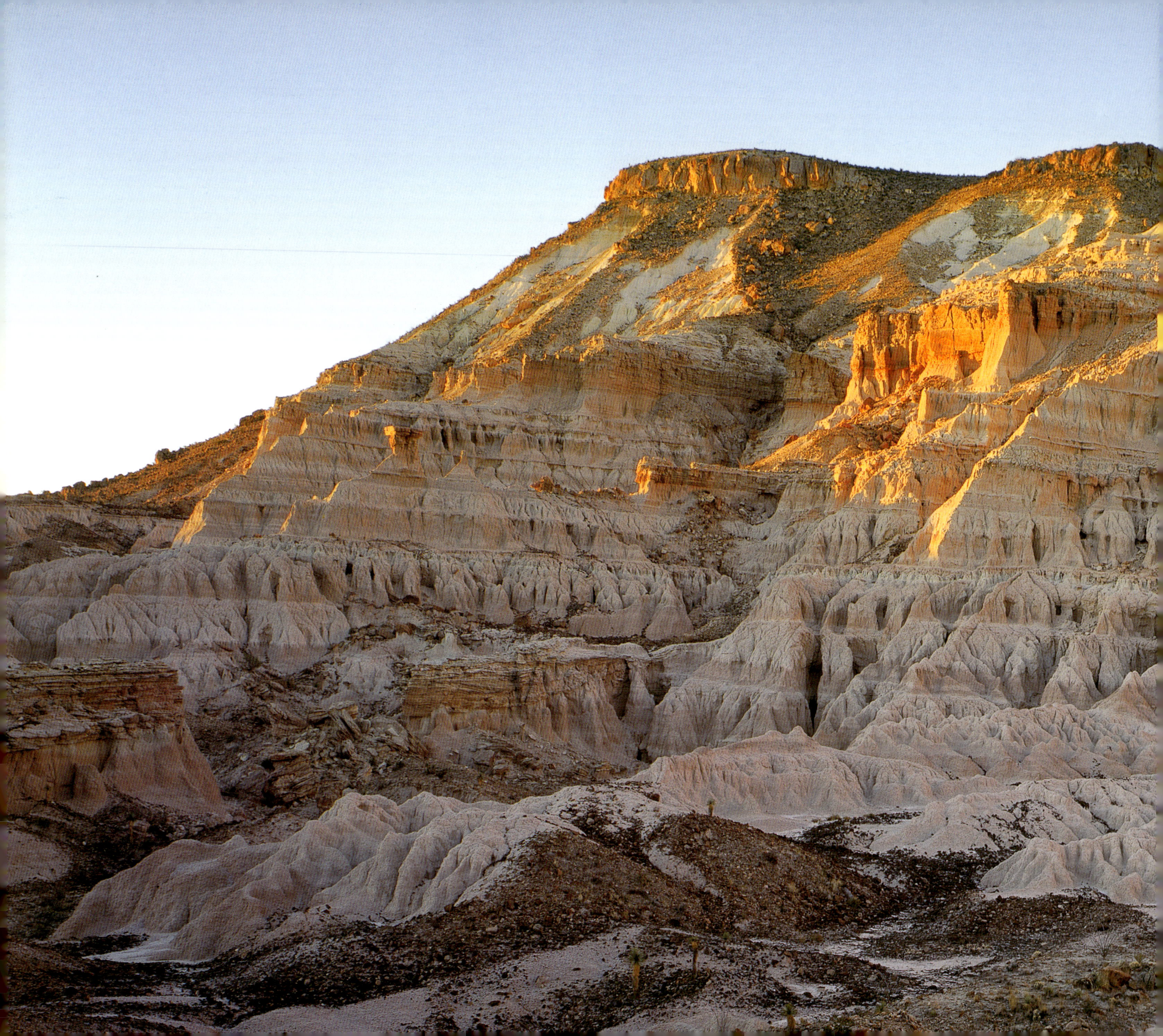

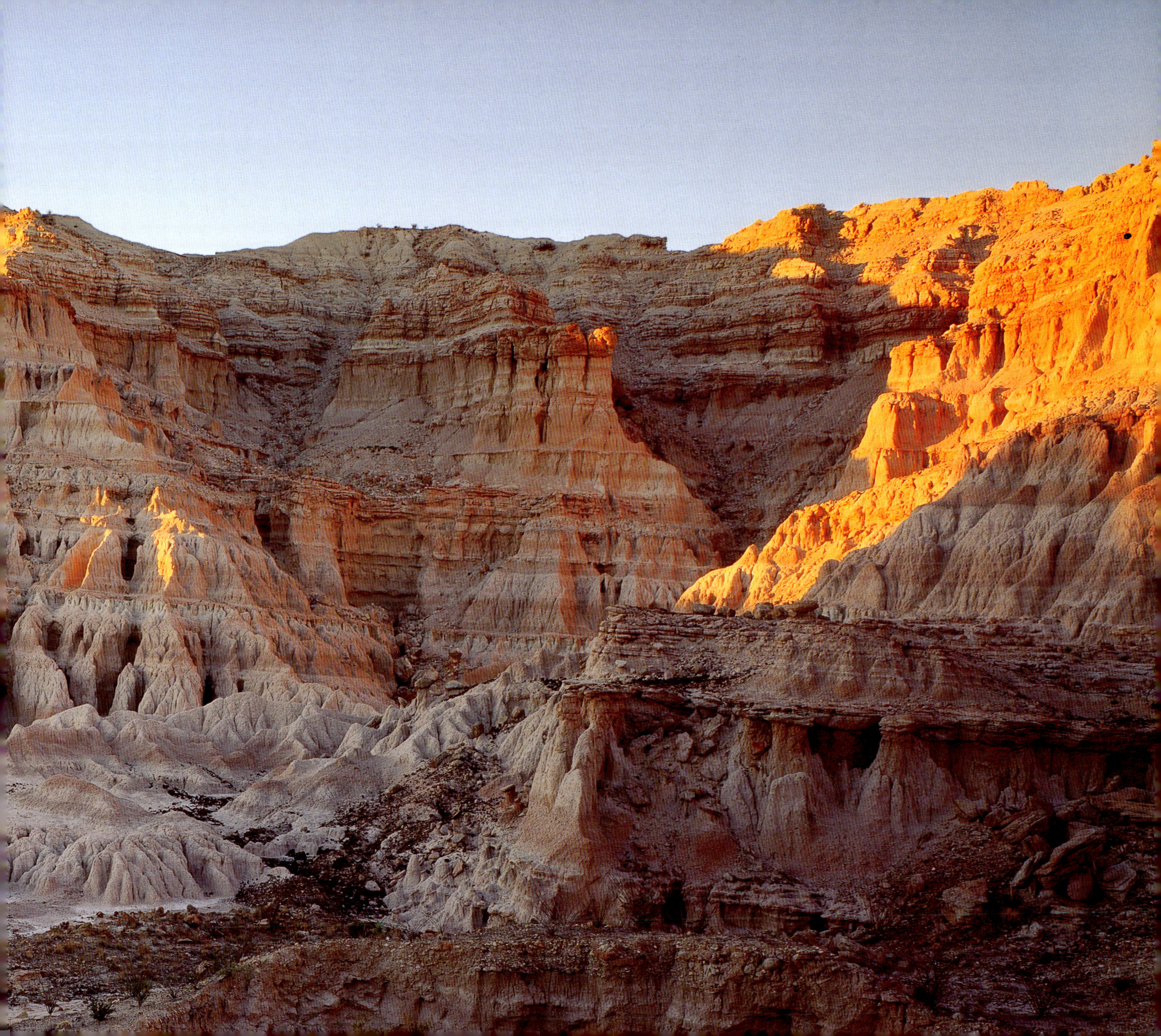

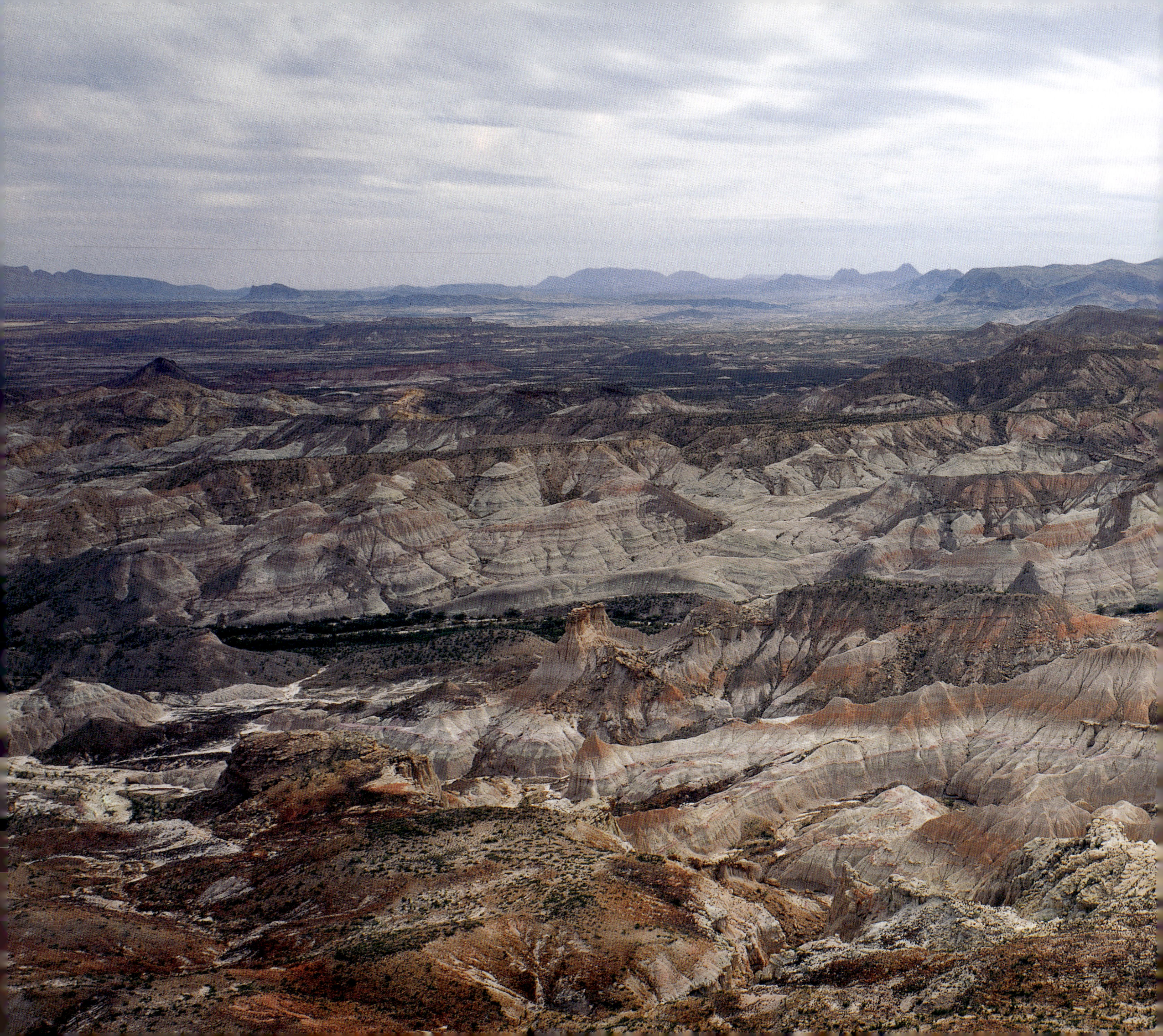

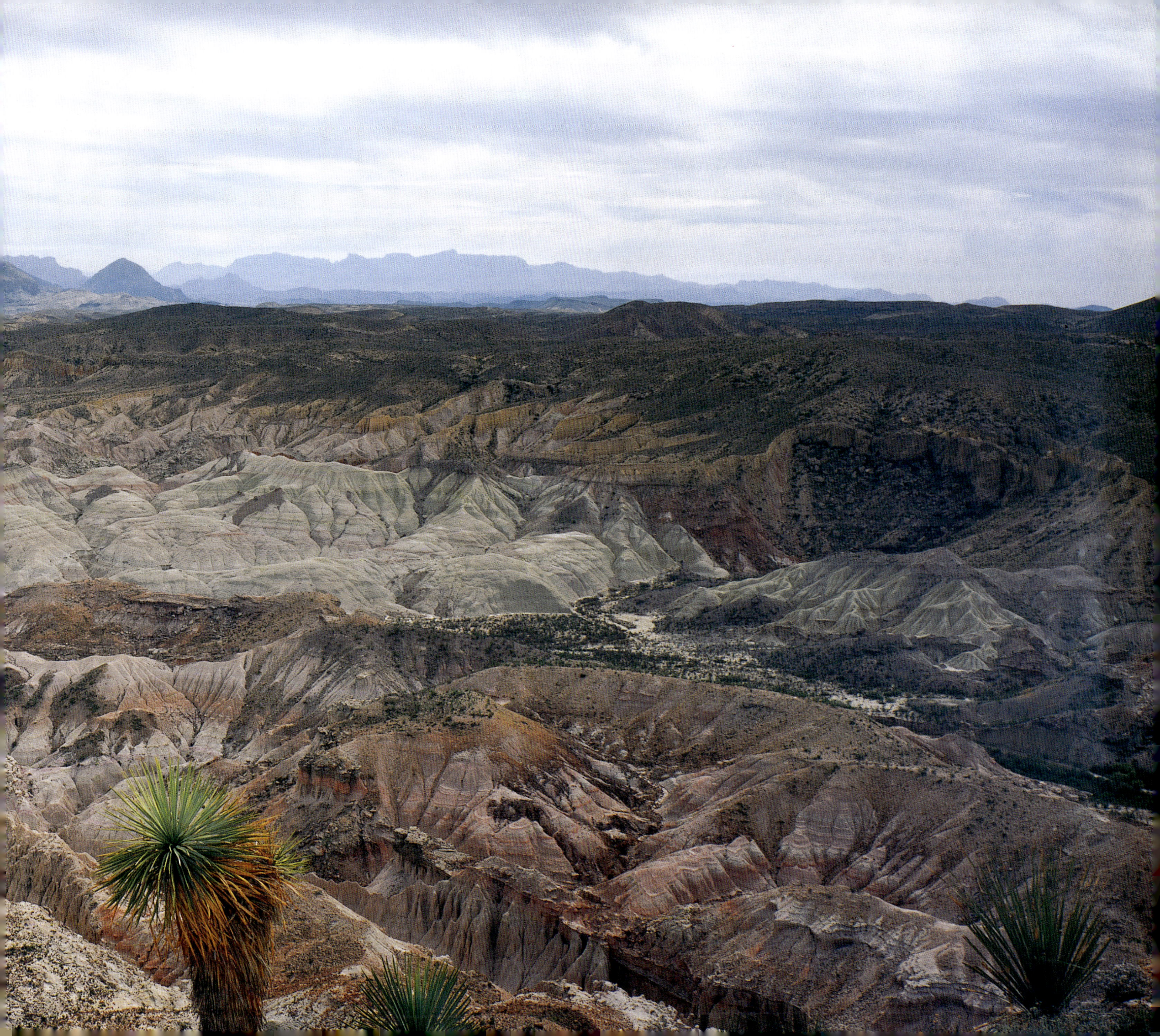

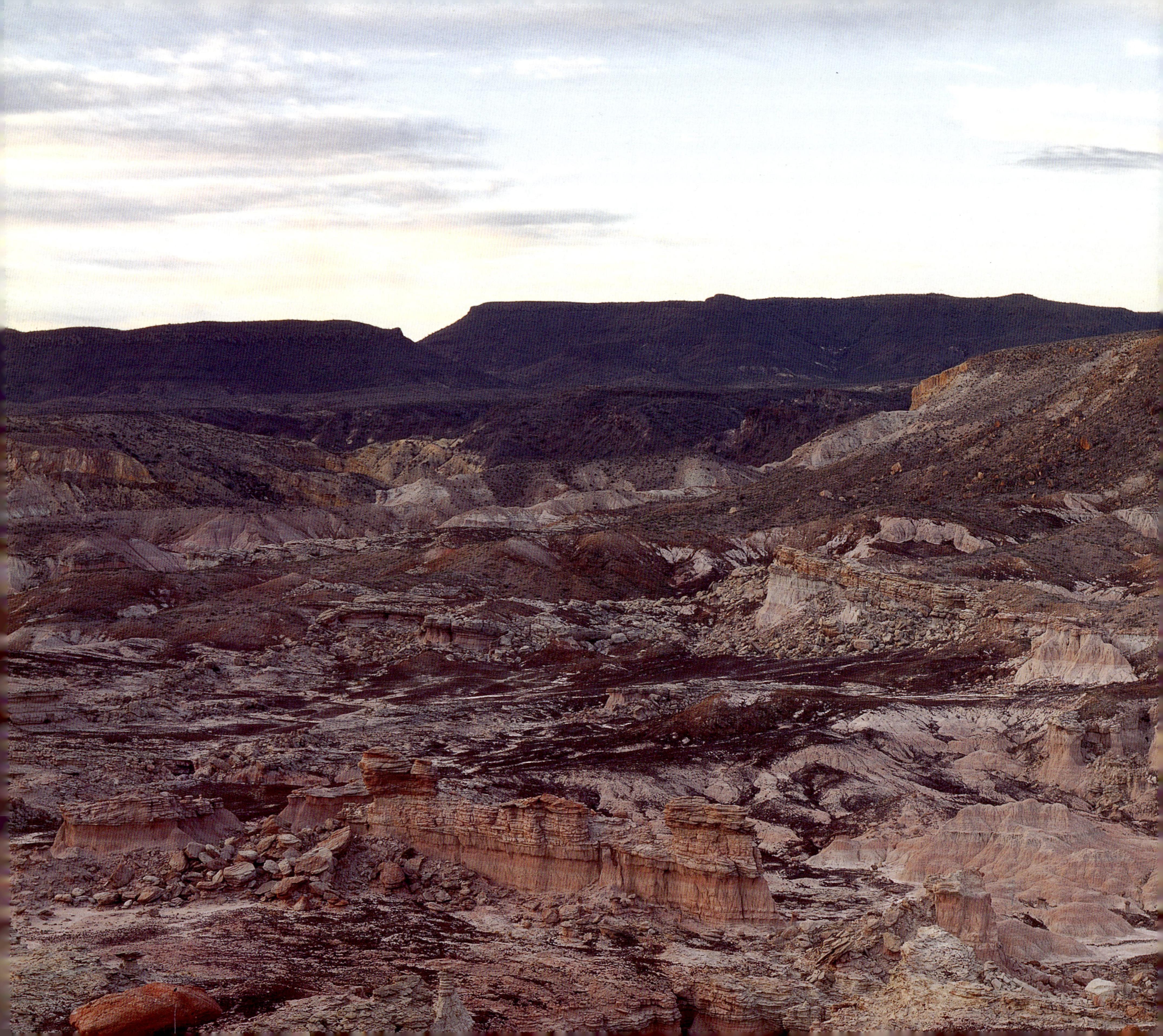

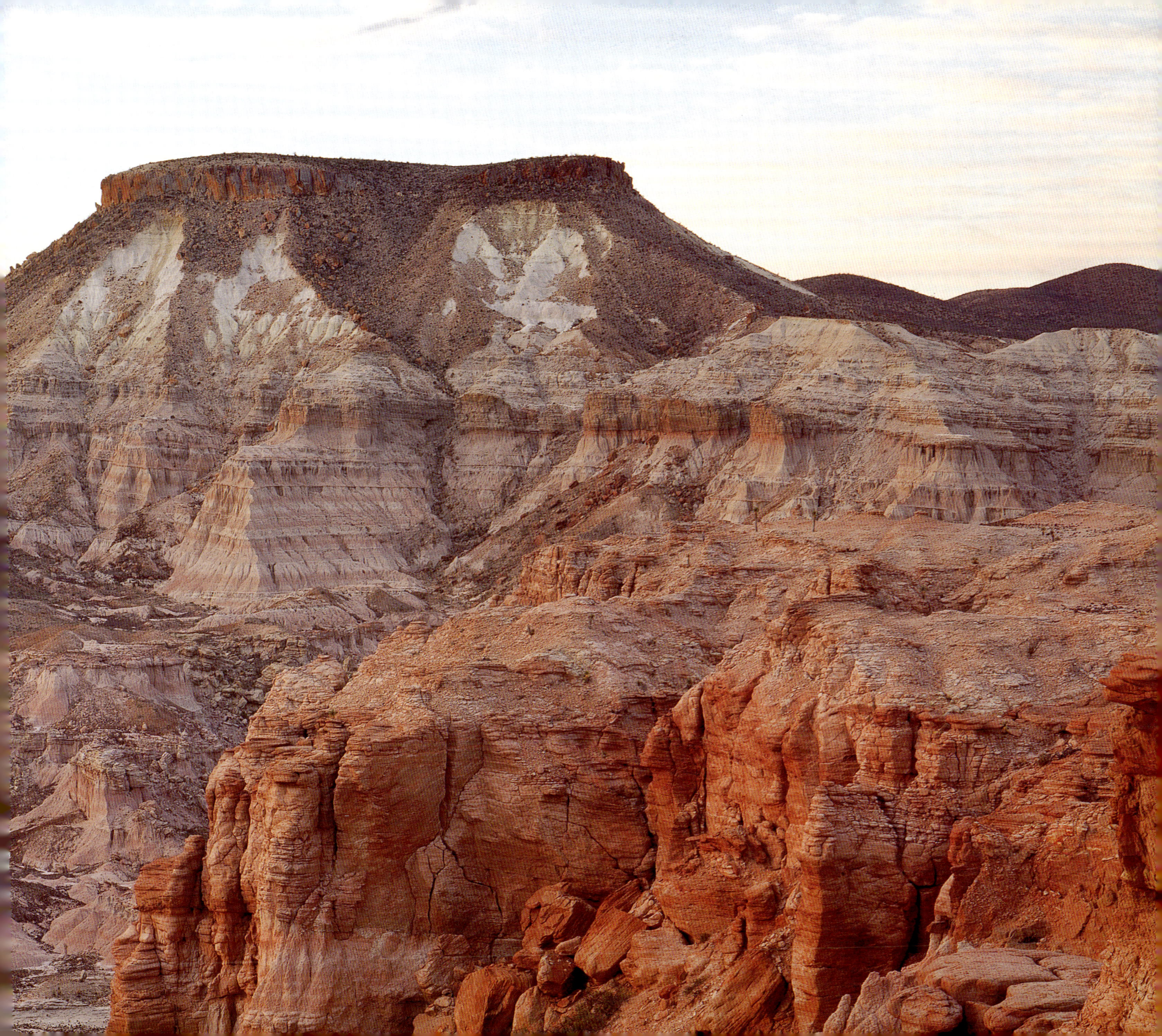

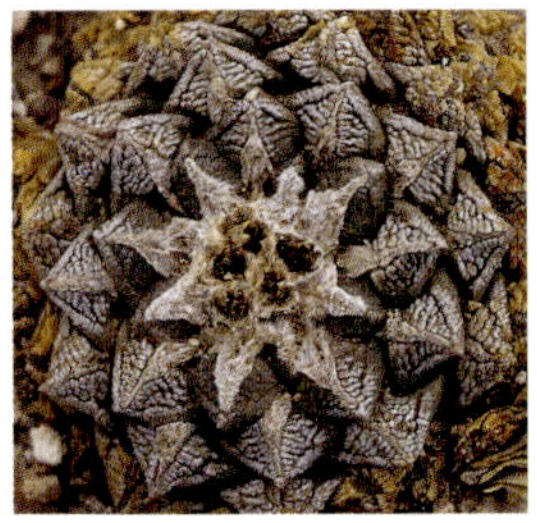